Readers Love Denise's Playful Math Books

"These games are great for using and practicing maths skills in a context in which there is some real motivation to do so. I love how they provide opportunities to explore a wide variety of approaches, including number bonds and logical thinking.

"My children are always pleased, even excited, when I suggest one of these games. Sometimes they even ask to play them unprompted!"
— Miranda Jubb, online reader review

"It revolutionized our homeschool this year."
— Caitlin Fitzpatrick Curley, My-Little-Poppies.com

"I have played several of these games with my son, and each one was met with delight on his part and the sharing of delightful conversation about numbers and thinking between us.

"I love what Gaskins has to say about working with your children as opposed to simply assigning them work to do. This sums up the philosophy that I try to keep forefront in our home.

"Highly, highly recommended."
— Amy, Hope Is the Word blog

"Wonderful games for students. The author includes a link for printable game boards, ensuring that I don't spend more time making games than playing them.

"Variations for each game = SO many ways to explore numbers.

"You will love this book."
— Marisa, online reader review

Books by Denise Gaskins

*Let's Play Math: How Families Can
Learn Math Together—and Enjoy It*

Math You Can Play Series
*Counting & Number Bonds
Addition & Subtraction
Math You Can Play Combo
Multiplication & Fractions
Prealgebra & Geometry*

Playful Math Singles Series
*Let's Play Math Sampler
70+ Things To Do with a Hundred Chart:
Word Problems from Literature
Word Problems from Literature Student Workbook
312 Things To Do with a Math Journal*

The Adventurous Student Journals Series

Printable Activity Guides for Teachers and Homeschoolers

MATH YOU CAN PLAY 4

Prealgebra & Geometry

Math Games for Middle School
Fourth to Ninth Grade

Denise Gaskins

Tabletop Academy Press

Copyright © 2021 Denise Gaskins
Print version 1.01
All rights reserved.
Except for brief quotations in critical articles or reviews, the purchaser or reader may not modify, copy, distribute, transmit, display perform, reproduce, publish, license, create derivative works from, transfer or sell any information contained in this book without the express written permission of Denise Gaskins or Tabletop Academy Press.

Tabletop Academy Press, Blue Mound, IL, USA
tabletopacademy.net

ISBN: 978-1-892083-46-3
Library of Congress Control Number: 2020914409

Disclaimer: This book is provided for general educational purposes. While the author has used her best efforts in preparing this book, Tabletop Academy Press makes no representation with respect to the accuracy or completeness of the contents, or about the suitability of the information contained herein for any purpose. All content is provided "as is" without warranty of any kind.

CREDITS:

Portions of this book were originally published on the Let's Play Math blog. DeniseGaskins.com

Cover photo by Betty_photo via iStock:
istockphoto.com/sg/photo/children-playing-cards-in-an-old-trailer-gm178363140-20748612

"Playground balance puzzle" drawing by Emma Baldwin.

Riffle shuffle photo by Johnny Blood (CC-BY-SA 2.0):
commons.wikimedia.org/wiki/File:Riffle_shuffle.jpg

Author photo by Christina Vernon:
melliru.com

Contents

Preface to the *Math You Can Play* Series vii

Acknowledgements ix

A Strategy for Learning 1

Teaching Math Through Play 5

Gather Your Game Supplies... 13

Prealgebra and Geometry Games 17

Number Properties 19

Integers 49

Operations and Functions 83

Geometry 141

Playing to Learn Math 185

More Than One Way to Solve It 187

Conclusion: Algebra Tells a Story 199

Resources and References 205

    Game-Playing Basics, From Set-Up to Endgame 206

    A Few of My Favorite Resources 212

    Quotes and Reference Links 217

    Index 231

    About the Author 241

Preface to the *Math You Can Play* Series

THE PLAYFUL, PUZZLE-SOLVING SIDE OF math has always attracted me. In elementary school, calculations were a tedious chore, but word problems provided the opportunity to try out my deductive powers. High school algebra and geometry were exercises in logical reasoning, and college physics was one story problem after another—great fun!

As my children grew, I wanted to share this attitude of mathematical play with them, but the mundane busyness of everyday life kept pushing aside my good intentions. Determined to make it happen, I found a way to defeat procrastination: invite friends to bring their kids over for a math playdate. We grappled with problems, solved puzzles, and shared games. Skeptical at first, the children soon looked forward to math club. When that gang moved on, their younger siblings came to play, and others after them. Sometimes we met weekly, sometimes monthly or just off and on. At our house, at the library, in the park—over two decades of playing math with kids.

Now I've gathered our favorite math club games into these *Math You Can Play* books. They are simple to learn, easy to set up, and quick to play, so even the busiest parents can build their children's mental math skills and promote logical thinking.

I hope you enjoy these games as much as we have. If you have any questions, I would love to hear from you.

— DENISE GASKINS
LETSPLAYMATH@GMAIL.COM

P.S.: If you've read the *Math You Can Play* books in order, you'll notice that I repeat myself in Section I and in the back-of-book reference section. I'm including the game set-up information and teaching tips in each book to make sure they can all stand on their own.

Besides, it's far too easy for us adults to fall back on the expectation that our children should learn the same way we were taught. Those "teaching math through play" tips in the first chapter are worth reading more than once.

Acknowledgements

No man is an island, entire of itself.
Every man is a piece of the continent, a part of the main.
If a clod be washed away by the sea,
Europe is the less, as well as if a promontory were.

— JOHN DONNE

Neither an island nor a promontory — I am that little clod supported by a continent of family, friends, and online acquaintances whose help and encouragement have made my math books possible. I can't express the debt I owe to my husband David, whose patience stretches far beyond what I deserve, or to our children, who taught me so much over the many years of homeschooling.

Thank you to Marilyn Kok and Sue Kunzeman, who brought their children to that first math playdate — and kept bringing them back. Special thanks to the many math club kids who joined in the activities and tested out the games.

Thank you to John Golden, whose Math Hombre blog inspires me to think deeply about math games, and to Sue VanHattum, whose work on her own book, *Playing with Math: Stories from Math Circles, Homeschoolers, and Passionate Teachers,* convinced me to bring my books back into print.

To my many thousands of book and blog readers, your comments have kept me going. To my fellow math bloggers, I've learned so much from you all!

Fervent thanks to my beta readers: Amber, Carla, Theresa, Casey (both of you!), Nicky, Becky, Sonya, Farrar, and Sue. And unending gratitude to my friend and editor Robin Netherton. Whatever mistakes remain are due to my continual tinkering with the text after it left her hands.

Section I

A Strategy for Learning

*You can think of puzzles and games
as the sugar that helps the medicine to go down,
and you're at least a bit healthier in your approach to math.*

*But even better than sugar and nasty medicine
is food that's delicious enough
to take away our craving for sugar
and nutritious enough
to take away any need for medicine.*

*Over time, perhaps
you can find the sweetness
in the math itself—
in a problem that inspires you
to work and struggle,
until you finally get it,
just for your own satisfaction.*

*Good problems can help us fall in love with math
and make a delicious meal of it,
sinking our teeth into tough problems,
tenderized by their intrigue.*

— Sue VanHattum

There should be no element of slavery in learning. Enforced exercise does no harm to the body, but enforced learning will not stay in the mind. So avoid compulsion, and let your children's lessons take the form of play.

— Plato

Teaching Math Through Play

IF A PERFECT TEACHER DEVELOPED the ideal teaching strategy, what would it be like?

- ♦ An ideal teaching strategy would have to be flexible, working in a variety of situations with students of all ages.
- ♦ It would promote true understanding and reasoning skills, not mere regurgitation of facts.
- ♦ It would prepare children to learn on their own.
- ♦ Surely the ideal teaching strategy would be enjoyable, perhaps even so much fun that the students don't realize they are learning.
- ♦ And it would be simple enough that imperfect teachers could use it, too.

This is idle speculation. No teaching strategy works with every student in every subject. But for math, at least, there is a wonderful way to stimulate our children's number and geometry skills while encouraging them to think: We can play games.

Math games push students to develop a creatively logical approach to solving problems. When children play games, they build reasoning skills that will help them throughout their lives. In the stress-free struggle of a game, players learn to analyze situations and draw conclusions. They must consider their options, change their plans in reaction to another player's moves, and look for the less obvious solutions in order to outwit their opponents.

Even more important, games help children learn to enjoy the challenge of thinking hard. Students willingly practice far more arithmetic than they would suffer through on a workbook page. Their vocabulary grows as they discuss options and strategies with their fellow players. Because their attention is focused on their next move, they don't notice how much they are learning.

And games are good medicine for math anxiety. Everyone knows it takes time to master the fine points of a game, so children can make mistakes or "get stuck" without losing face.

If your child feels discouraged or has an "I can't do it" attitude toward math, try taking him off the textbooks for a while. Feed him a strict diet of games, and his eyes will soon regain their sparkle. Children love beating a parent at a math game. And if you're like me, your kids will win more often than you'll want to admit.

Math You Can Play

Clear off a table, find a deck of cards, and you're ready to enjoy some math. Most of the games in this book take only a few minutes to play, so they fit into your most hectic days.

In three decades of teaching, I've noticed that flexibility with mental calculation is one of the best predictors of success in high school math and beyond. So the *Math You Can Play* games stretch your children's ability to manipulate numbers in their heads. But unlike the typical "computerized flash card" games online, most of these encourage your children to think strategically, to compare different options in choosing their moves.

"Be careful! There are a lot of useless games out there," says math professor and blogger John Golden. "Look for problem solving, the need for strategy, and math content.

"The best games offer equal opportunity (or nearly so) to all your students. Games that require computational speed to be successful will disenfranchise instead of engage your students who need the game the most."

Each book in the *Math You Can Play* series features twenty or more of my favorite math games, offering a variety of challenges for all ages. If you are a parent, these games provide opportunities to enjoy quality time with your children. If you are a classroom teacher, use the games as warm-ups and learning center activities or for a relaxing review day at the end of a term. If you are a tutor or homeschooler, make games a regular feature in your lesson plans to build your students' mental math skills.

Know that my division of these games by grade level is inherently arbitrary. Children may eagerly play a game with advanced concepts if the fun of the challenge outweighs the work involved. Second- or third-grade students can enjoy some of the games in the prealgebra book. On the other hand, don't worry that a game is too easy for your students, as long as they find it interesting. Even college students enjoy a round of Farkle (in the addition book) or Wild and Crazy Eights (a childhood classic from the counting book). An easy game lets the players focus most of their attention on the logic of strategy.

As Peggy Kaye, author of *Games for Math*, writes: "Children learn more math and enjoy math more if they play games that are a little too easy rather than a little too hard."

Games give children a meaningful context in which to ponder and manipulate numbers, shapes, and patterns, so they help players of all skill levels learn together. As children play, they exchange ideas and insights.

"Games can allow children to operate at different levels of thinking and to learn from each other," says education researcher Jenni Way. "In a group of children playing a game, one child might be encountering a concept for the first time, another may be developing his/her understanding of the concept, a third consolidating previously learned concepts."

Talk with Your Kids

The modern world is a slave to busyness. Marketers tempt well-intentioned parents with toys and apps that claim to build academic skills while they keep our children occupied. Homeschoolers dream of finding a curriculum that lets the kids teach themselves. And even the most attentive classroom teachers may hope that game time will give them a chance to correct papers or catch up on lesson plans.

Be warned: Although children can play these games on their own, they learn much more when we adults play along.

When adults join the game, we reinforce the value of mathematical play. By giving up our time, we prove that we consider this just as important as *[insert whatever we would have been doing]*. If a game is worthy of our attention, then it becomes more attractive to our children.

Also, as we watch our kids' responses and listen to their comments during play, we discover what they understand about math. Where do they get confused? What do they do when they're stuck? Can they use the number patterns they've mastered to figure out something they don't know? How easily do they give up?

"Language should be part of the activity," says math teacher and author Claudia Zaslavsky. "*Talk* while you and your child are playing games. Ask questions that encourage your child to describe her actions and explain her conclusions."

Real education, learning that sticks for a lifetime, comes through person-to-person interactions. Our children absorb more from the give and take of discussion with an adult than from even the best workbook or teaching video.

If you're not sure how to start a conversation, browse the stories at Christopher Danielson's Talking Math with Your Kids blog.[†]

---

† *talkingmathwithkids.com*

As homeschooler Lucinda Leo explains, "With any curriculum there is the temptation to leave a child to get on with the set number of pages while you get on with something else. My long-term goal is for my kids to be independent learners, but the best way for that to happen is for me to be by their side now, enjoying puzzles and stories, asking good questions and modelling creative problem-solving strategies."

And playing math games.

Mix It Up

Games evolve as they move from one person to another. Where possible, I have credited each game's inventor and told a bit of its history. If you want more information, look up that person the "Quotes and Reference Links" appendix.

But some games have been around so long they are impossible to trace. Many are adapted from traditional childhood favorites. For example, I was playing Tens Concentration with my math club kids years before I read about it in Constance Kamii's *Young Children Reinvent Arithmetic.* Similarly, an uncountable number of parents and teachers have played Math War with their students; a few of my variations are original, but the underlying idea is far from new.

Sometimes, as in the Math War variations, the basic rules of play stay the same, making the new games easy for children to learn. Likewise, we can adapt the multiplication classic Product Game to practice fractions, decimals, or basic algebra without changing the rules, simply by switching the gameboard.

In other cases, we change the rules themselves to make something original. Consider the lineage of Forty-Niners, featured in the *Math You Can Play* addition book. First someone invented dice, and generations of players created a multitude of folk games, culminating in Pig. Using cards instead of dice and adding a Wild West theme, James Ernest created the Gold Digger version and gave it away at his website. Teachers wanted their students to practice with bigger numbers, so they tried a regular deck of playing cards, and the game became Stop or Dare at the Nrich Maths website. For my variation, I increased the risk level by turning all the face cards into bandits and adding the jokers as claim jumpers.

Game rules are a social convention, easy to change by agreement among the players. Feel free to invent your own rules, and encourage your children to modify the games as they play. For instance:

- ◆ Can you make the game easier, so young children can play? Or harder, to challenge adults?
- ◆ What if you changed the number of cards you draw, or how many dice you throw?
- ◆ Can you invent a story to explain the game—like James Ernest did with Gold Digger—or tie it to a favorite book?
- ◆ If the game uses cards, can you figure out a way to play it with dice or dominoes? Or transfer it to a gameboard?
- ◆ If the game uses a number chart, could you play it on a clock or calendar instead? Or is there a way to use money in the game?
- ◆ Or can you change it into a whole-body action game? Perhaps using sidewalk chalk?

As children tinker with the game, it prompts them to think more deeply about the math behind it.

Unschooling advocate Pam Sorooshian explains the connection between games and math this way:

> "Mathematicians don't sit around doing the kind of math that you learned in school. What they do is 'play around' with number games, spatial puzzles, strategy, and logic.
>
> "They don't just play the same old games, though. They change the rules a little, and then they look at how the game changes.
>
> "So, when you play games, you are doing exactly what mathematicians really do—IF you fool with the games a bit, experiment, see how the play changes if you change a rule here

and there. Oh, and when you make up games and they flop, be sure to examine why they flop—that is a big huge part of what mathematicians do, too."

Teaching Tips

Don't work through the games in this book in order. Each chapter is arranged by difficulty level, with the most challenging games at the end. The first game in Chapter 5, for example, will be much easier for your students than a game near the end of Chapter 3. As you read, you may want to place a bookmark in each chapter at the game that feels right for your kids.

Try to let children learn by playing. Explain the rules as simply as possible and jump into the fun. You can add details, exceptions, and special situations as they come up during play or before starting future games. At our house, we play a couple of practice rounds first, and I make sure I have explained all the rules before we keep score.

Card games have a traditional ethic that guides players in choosing who gets to deal, who goes first, what to do if something goes wrong in the deal or during play, and more. If you are unsure about questions of this sort, read "Game-Playing Basics" in the appendix.

Many of the game listings include suggestions for *house rules,* which are optional modifications. The way a game is played varies from one place to another, and only a few tournament-style games have an official governing body. If you're not playing in an official competition, then everything is negotiable. Players should make sure they agree on the rules before starting to play.

Finally, although the point of these games is for children to practice mental math, please don't treat them as worksheets in disguise. A game should be voluntary and fun. No matter how good it sounds to you, if a game doesn't interest your kids, put it away. You can always try another one tomorrow.

You'll know when you find the perfect game because your children will wear you out wanting to play it again and again and again.

You can download free printable versions of all the gameboards in this book. Look for the *Prealgebra & Geometry Printables* companion file on my Tabletop Academy Press publishing site: *tabletopacademy.net/free-printables*

We do not stop playing because we grow old. We grow old because we stop playing.
— ANONYMOUS

Gather Your Game Supplies

I HAVE A LIMITED AMOUNT of free time, and I don't want to spend it cutting out specialized game pieces or cards. A few games require printable cards or gameboards, but most of the games in the *Math You Can Play* series use basic items you already have, such as playing cards and dice.

A Deck of Cards

Whenever a game calls for playing cards, I use an international standard poker- or bridge-style *deck* (or *pack*; the terms are interchangeable). There are fifty-two cards in four suits—spades (the pointy black shape), hearts, clubs (the clover shape), and diamonds—with thirteen cards per suit. The number cards range from the ace to ten, and each suit has three face cards called jack, queen, and king.

Your deck may have one or two additional cards called jokers, which are not officially part of the deck but may be used for some games. You may also allow jokers to be *wild cards*—taking the value of any card that the player chooses.

If I'm playing with younger children, we use the forty number cards (ace through ten in all four suits) from a standard deck. The ace counts as one. With older children, we may count all face cards as ten or go for larger numbers: jack = eleven, queen = twelve, and king = thirteen. Sometimes I give players a choice to let the ace count as one or "king plus one"—so if the king counts as ten, players may score the ace as either one or eleven, whichever works best for their hand. In a few games, we use the queens as zeros, because the Q is round enough for pretend.

You can play with whatever card decks you have on hand. Uno cards are numbered zero to nine, Phase 10 cards have one to twelve, and Rook cards go from one to fourteen. Rummikub tiles use the numbers one to thirteen. You could adapt most games to any of these.

Gameboards

For many games, hand-drawn gameboards work fine. One reason Tic-Tac-Toe is a perennial favorite is that children can draw the board whenever they want to play.

Some games use graph paper as a gameboard. As students grow

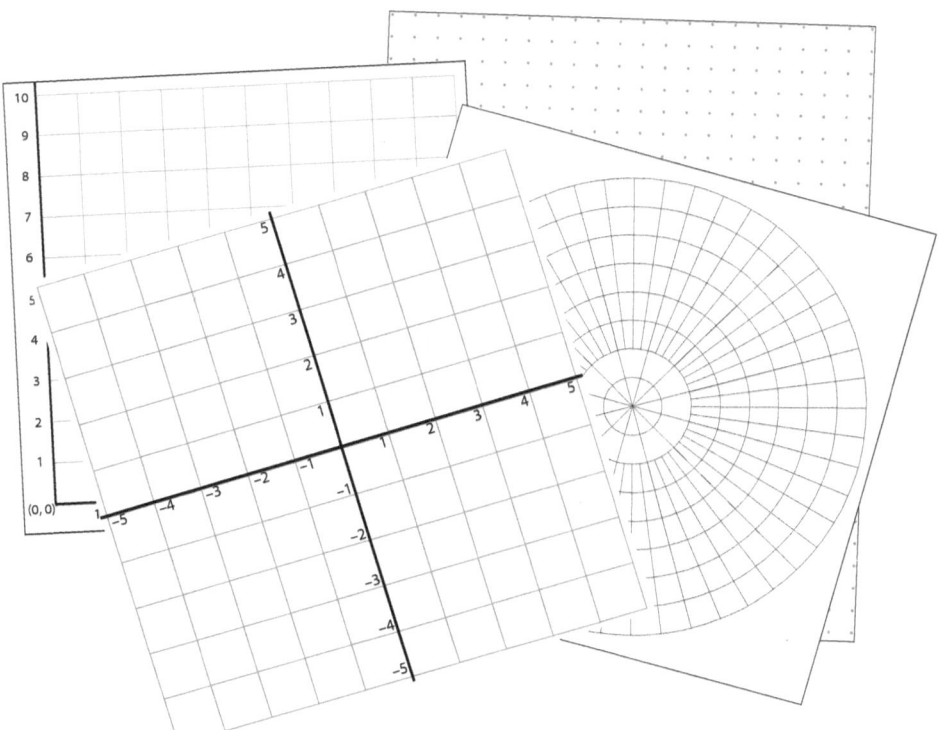

toward algebra and geometry, they need to master the *coordinate plane*—the square grid on which we graph lines and equations. So we have several games to play with coordinate graphing.

I've created a free PDF packet of graph paper and other gameboards—the *Prealgebra & Geometry Printables*—which you can download from my publisher's website. You may reproduce these for use within your own family, classroom, or homeschool group.[†]

To save printer ink, you may want to reuse gameboards. Print the gameboard on cardstock and laminate it—I love my laminator!—or slip the printed board into a clear (not frosted) page protector, adding a few sheets of card stock or the back of an old notebook for stiffness. Then your children can mark moves with dry-erase markers and wipe them clean with an old, dry cloth.

Other Bits and Pieces

Many games call for small toy figures or other items to mark each player's position or moves. If two different types of tokens are needed, you may borrow the pieces from a checkers game. Or use pennies and nickels, milk jug lids in different colors, dried pinto and navy beans, or inexpensive glass gems from the craft section of your local department store. Or let players create their own game pieces by cutting small shapes from colored construction paper, decorating if desired.

For some games, you can replace tokens with colored erasable markers on a laminated gameboard. The colors must be different enough to distinguish each player's moves. Or players can use different symbols, like the X and O of Tic-Tac-Toe. Some colored dry-erase markers leave stains, but usually you can wash off stubborn marks with rubbing alcohol or window cleaner.

When a game calls for dice, I have in mind the normal six-sided cubes with dots for the numbers. Most games only need one or two dice, but Farkle requires six. In many of the games, you may substitute higher-numbered dice for a greater challenge. And children enjoy

---

[†] *tabletopacademy.net/free-printables*

using novelty dice when making up their own games.

A few games call for either a double-six or double-nine set of dominoes. If you are buying these, I recommend getting the larger set. You can always set aside the higher-numbered tiles when playing with young children.

Ready to Play?

If you want to put together a game box to keep all your supplies in one place, here's a checklist:

- ◆ Standard playing cards (two or more decks)
- ◆ Pencils or pens
- ◆ Colored felt-tip markers or colored pencils
- ◆ Blank paper
- ◆ At least two kinds of tokens
- ◆ Several dice
- ◆ Dominoes
- ◆ Ruler or other straightedge
- ◆ Assorted graph paper
- ◆ Printed gameboards from the free download file *Prealgebra & Geometry Printables*
- ◆ Erasable markers for use on laminated gameboards, and a cloth or paper towel for wiping them clean (optional)

Section II

Prealgebra and Geometry Games

Number Properties

*Of all the myths about mathematics,
the one I find most blatantly wrong
is the idea that some people
are just born knowing the answers.*

*In my experience,
when you confront a genuine puzzle,
you start out not knowing,
no matter who you are.*

*Moreover, 'knowing' the answers can be a trap;
learning mathematics is about
looking at what you thought you understood
and seeing that there's deeper mystery there
than you realised.*

— Dan Finkel

Activity: Polite Numbers

PREALGEBRA MARKS A TRANSITION BETWEEN the basic arithmetic of grade school and the more abstract topics studied in high school. Students review and consolidate their understanding of standard calculations, dip their toes into the waters of algebraic thinking, and—if they're lucky—wander down a few mathematical rabbit trails.

Number theory studies the properties of whole numbers and their relationships to one another. One of the greatest mathematicians of all time once said:

> *"Mathematics is the queen of the sciences, and number theory is the queen of mathematics."*
> —CARL FRIEDRICH GAUSS

Let's follow that rabbit down a number theory trail...

Did you know that numbers can be polite? In math, a *polite number* is any number we can write as the sum of two or more consecutive natural numbers.

Consecutive means numbers that come one right after another in the counting sequence. And *natural numbers* are the positive counting numbers 1, 2, 3, and so on—the numbers young children naturally learn first. Sometimes mathematicians include zero in the natural numbers, but not for this puzzle.

Here are a few examples. Five is a polite number, because we can write it as the sum of two consecutive natural numbers:

$$5 = 2 + 3$$

Nine is a doubly polite number, because we can write it two ways:

$$9 = 4 + 5$$
$$9 = 2 + 3 + 4$$

And fifteen is an amazingly polite number. We can write fifteen as the sum of consecutive numbers in three ways:

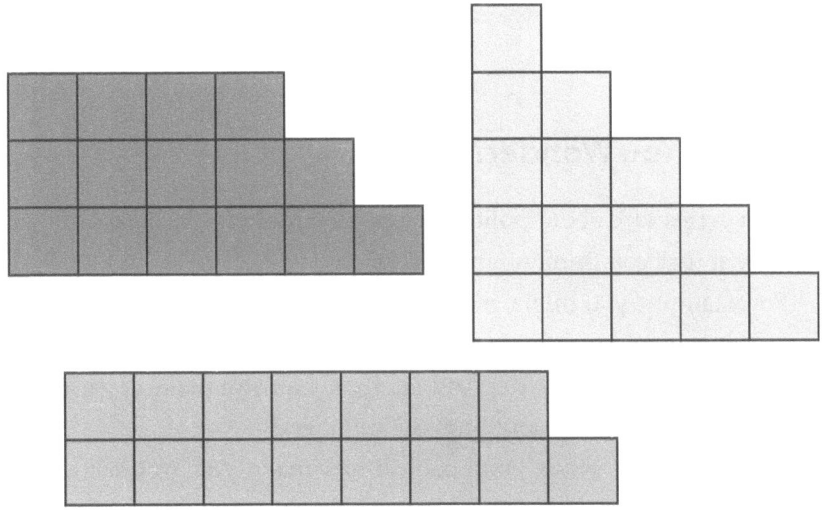

We can also call these *staircase numbers* because we can arrange that many blocks into steps. Fifteen blocks can make a stairway in three different ways.

$$15 = 7 + 8$$
$$15 = 4 + 5 + 6$$
$$15 = 1 + 2 + 3 + 4 + 5$$

How many other polite numbers can you find?

What Do You Notice?

Are all numbers polite? Or can you find an impolite number?

An *impolite number* is one that cannot be written as the sum of two or more consecutive natural numbers. It's impossible to make into block stairs. Impolite numbers are also known as "The Unsummables."

Make a collection of polite and impolite numbers. Find as many as you can. Count the different ways you can write each polite number as a sum of consecutive natural numbers.

How can you check whether you've found all the possibilities? Can you think of a way to organize your collection so you can look for patterns?

What Do You Wonder?

Make a conjecture about polite or impolite numbers. A *conjecture* is a statement that you think might be true.

For example, you might make a conjecture that "All odd numbers are…" How would *you* finish that sentence?

Make another conjecture. And another. Can you make at least five conjectures about polite and impolite numbers?

Does thinking about your conjectures make you wonder about math?

Can you think of any way to test your conjectures, to discover if they will always be true?

Real-Life Mathematics

This is how mathematics works. Mathematicians notice something interesting about certain numbers, shapes, or ideas. They play around and explore how those relate to other ideas. After collecting a set of interesting things, they think about ways to organize them. They wonder about patterns and connections. They make conjectures and try to imagine ways to test them.

And mathematicians talk with one another and compare their ideas. In real life, math is a very social game.

So experiment with polite and impolite numbers. Share your conjectures with friends, and see what they think. Or if you don't have friends who likes to play with math, compare your ideas with the list of student conjectures at the Ramblings of a Math Mom blog.[†]

[†] *mathmomblog.wordpress.com/2007/10/12/middle-schoolers-and-the-unsummables*

Factors and Multiples

MATH CONCEPTS: multiplication, division, factors and multiples.
PLAYERS: two to four.
EQUIPMENT: printed hundred chart, pencils or markers. Calculator optional.

How to Play

Print a single hundred chart for the players to share.

The first player chooses an even number less than fifty and marks that square on the hundred chart by coloring the square or drawing a large X over the number. Then the next player marks one factor or multiple of that number.

Players alternate turns, each marking an available (unmarked) factor or multiple of the last number played.

The player who takes the last legal number, leaving the other players with no move, wins the game.

1	2	3	4	5	6	7	8	9	10
11	12	13	14	15	16	17	18	19	20
21	22	23	24	25	26	27	28	29	30
31	32	33	34	35	36	37	38	39	40
41	42	43	44	45	46	47	48	49	50
51	52	53	54	55	56	57	58	59	60
61	62	63	64	65	66	67	68	69	70
71	72	73	74	75	76	77	78	79	80
81	82	83	84	85	86	87	88	89	90
91	92	93	94	95	96	97	98	99	100

91	92	93	94	95	96	97	98	99	100
81	82	83	84	85	86	87	88	89	90
71	72	73	74	75	76	77	78	79	80
61	62	63	64	65	66	67	68	69	70
51	52	53	54	55	56	57	58	59	60
41	42	43	44	45	46	47	48	49	50
31	32	33	34	35	36	37	38	39	40
21	22	23	24	25	26	27	28	29	30
11	12	13	14	15	16	17	18	19	20
1	2	3	4	5	6	7	8	9	10

You can play many games on a hundred chart.
The free *Prealgebra & Geometry Printables* download file
includes both top-down and bottom-up versions.

Variations

For a shorter game, cross out rows to limit the playable area of the gameboard. Try playing with 1–30 or 1–60.

FACTORS AND MULTIPLES COOPERATIVE GAME: Work together to find a long set of factors and multiples. Try to keep the game going, not to block the next player. Keep a list of your numbers in the order you use them. Can you mark more than half of the chart without breaking the chain?

History (and a Puzzle)

Factors and Multiples comes from the Nrich Maths website, a great place to visit and explore. You'll find an amazing variety of mathematical puzzles, games, and challenge problems for all ages.

PUZZLE: Why is there a restriction on the first number marked? Try playing a two-player game without that rule and see what happens. Can you figure out a strategy for the first player to guarantee a win?

Words to Know

The words *factor* and *multiple* always refer to whole numbers, not fractions or decimals. *Factors* are the numbers you multiply, and *multiples* are the answers you get. A factor always divides evenly (without remainder) into any of its multiples—so a factor may also be called a *divisor* of its multiple.

For example, the factors of twelve are 1, 2, 3, 4, 6, and 12, because those are all the whole numbers you can multiply to get twelve:

$$1 \times 12 = 12$$
$$2 \times 6 = 12$$
$$3 \times 4 = 12$$

And twelve is a multiple of each of its factors.

Words to Know

A *prime number* is a natural number with exactly two factors: one and itself. A *composite number* has more than two factors. By definition, the number one is neither prime nor composite.

Tax Collector

MATH CONCEPTS: multiplication, prime numbers, factors and multiples.
PLAYERS: two or more players or two teams.
EQUIPMENT: blank paper or a printed hundred chart, pencils or markers. Calculator optional.

Set-Up

Choose a set of numbers (the more numbers, the longer the game) such as 1–30 or 1–100. Write the numbers at the top of your paper, so you can mark them out as they are used. Or use a printed hundred chart. Keep score in the margins or on scratch paper.

How to Play

One player is the tax collector, and all the others work together as a team of shop owners. Or the players on one team are tax collectors, and the other team entrepreneurs.

First, a shop owner marks any available number by crossing it out or coloring that square on the chart. Add this number to the running total of shop owner profits.

Once a number is marked by any player, it cannot be used again.

After each shop-owner turn, the tax collector marks all the factors of that number that have not been previously scored. Add these numbers to the running total of taxes.

Continue turns with shop owners claiming a profit amount and the tax collector looking for factors to tax. The game ends when the shop owners have no legal play—that is, there are no numbers with factors left. The tax collector claims all remaining numbers.

Whoever collects the most money (profit or tax) wins.

WARNING: You must always pay the tax collector! No shop owner may mark a number that doesn't have any factors remaining. If you try to claim a number with no factors, the tax collector takes that whole amount as a penalty.

Variations

THE FACTOR GAME: Alternate roles on each turn. The first player chooses a number, and the second collects the tax. Then the second player chooses a number, and the first one collects the tax. At the end of the game, neither player gets the unclaimed numbers.

FACTOR BLASTER: Players alternate roles, as in the Factor Game, and they may choose any unmarked number—including numbers *without* factors for the other player to claim. In Tax Collector or the Factor Game, you may choose only one prime number (you'll find out why as you play), but the lack of penalties in Factor Blaster makes the prime numbers into prime targets.

History

In *Simply Great Math Activities: Number and Operations,* authors Bill Lombard and Brad Fulton write:

> *"The tax collector usually wins the first game so handily that the students are left believing there is no way for them to win this game. However, assure them there are many ways for the taxpayer to win, and ask them to play another game."*

Teacher coach Terry Kawas created the Factor Blaster variation that emphasizes the importance of prime numbers, to prepare for teaching prime factoring.

Great Escape I

MATH CONCEPTS: multiplication, division, factors and multiples, odd and even, prime numbers, square numbers, cubic numbers.
PLAYERS: two to four.
EQUIPMENT: printed hundred chart, playing cards (with jokers), different colored pencils or markers.

Set-Up

You are a flight lieutenant in the Royal Air Force, captured by the Axis powers during World War II and held in a prisoner-of-war camp. Your mission is to dig an escape tunnel from your barracks to the neighboring woods.

Print a single hundred chart for players to share. Each player needs a distinctly colored pencil or marker.

Spread the playing cards face down on the table as a fishing pond.

How to Play

Players take turns drawing a card from the fishing pond and coloring a matching number square on the hundred chart, according to the following code:

- Ace = any odd number
- 2 = any even number
- 3–10 = multiple of the card number
- Jack = prime number
- Queen = square number
- King = cubic number
- Joker = multiple of one *(All* whole numbers are a multiple of one, so jokers are wild cards.)

Words to Know

Square numbers are numbers that, if you arrange that many items in an array of rows and columns, can create a perfect square. For example, three rows with three items each create a square with nine items, so nine is a square number. Every square number is the product of the two sides of its square shape:

One row of one item = 1 × 1 = 1
Two rows of two items = 2 × 2 = 4
3 × 3 = 9, etc.

Cubic numbers (also called *cube numbers* or *perfect cubes*) are made with layers of perfect squares to form a cube. Three layers, each with three rows of three, create a cube of twenty-seven items. So twenty-seven is a cubic number. To calculate a cubic number, multiply the square number times the number of layers:

(One row of one) × 1 = 1
(Two rows of two) × 2 = 8
(3 × 3) × 3 = 27, etc.

Each player's first square must be on a vertical or horizontal outer edge of the gameboard. Circle that starting position with your color. After that, you may color in any unclaimed square.

If there is no number square that matches your card, you lose that turn. For instance, there are only four cubic numbers on the hundred chart. After those are all claimed, kings become "lose a turn" cards.

At the end of your turn, mix your card back into the fishing pond.

Your goal is to complete a connected path of squares (meeting at sides or corners) from your starting position to the opposite side of the gameboard. The first player to finish such a path escapes from the prison camp and wins the game.

Variations

For a more difficult challenge, players must tunnel directly from one square to the next. Each number you mark must touch one of your previously colored squares at a side or corner.

Or for a simpler game with no blocking, give each player a separate hundred chart gameboard. In this variation, tunnel squares must be attached at their sides; squares meeting only at a corner do not count as connected. To speed up the game, all players may draw cards and choose squares at the same time.

History

The Great Escape by Australian journalist and fighter pilot Paul Brickhill is a true story of perseverance, heroism, and tragedy during World War II. This game comes from Marina Singh's MathCurious blog.

Great Escape II

MATH CONCEPTS: multiplication, division, factors and multiples, odd and even, prime numbers, square numbers, cubic numbers.
PLAYERS: two to four (a cooperative game).
EQUIPMENT: printed hundred chart, playing cards (with jokers), pencils or markers.

Set-Up

You are a flight lieutenant in the Royal Air Force, captured by the Axis powers during World War II and held in a prisoner-of-war camp. You're working with your fellow prisoners to escape before the Nazi guards discover your tunnel.

Print a single hundred chart for players to share. Spread the playing cards face down on the table as a fishing pond.

How to Play

First, place the guard towers: Each player draws two number cards. Tens count as zero for this part of the game. If you get a face card or joker, draw again. Arrange your two cards to make a two-digit number. For example, if the cards are 3 and 8, you could make 38 or 83. Place a guard tower on that square by drawing an X with a circle around it. Mix all the cards back into the fishing pond.

Then players take turns drawing two cards from the fishing pond and marking two number squares — one for the escape tunnel and one for a patrolling Nazi guard — according to the following code:

- ♦ Ace = any odd number
- ♦ 2 = any even number
- ♦ 3–10 = multiple of the card number
- ♦ Jack = prime number

- Queen = square number
- King = cubic number
- Joker = multiple of one *(All* whole numbers are a multiple of one, so jokers are wild cards.)

Choose which of your two cards to use for the escape tunnel and color in a square that matches it. Then find a square that matches the other card and mark it with an X to represent the guard. Once a square is marked, it cannot be used in future turns.

Each player's initial escape tunnel square must be on a vertical or horizontal outer edge of the gameboard. The first player may choose any edge. The second player must start on the opposite side of the board from the first. The third chooses either unmarked edge, and the fourth player takes what's left.

If you can't find a square on your edge that matches either of your cards, you must still mark a Nazi guard position.

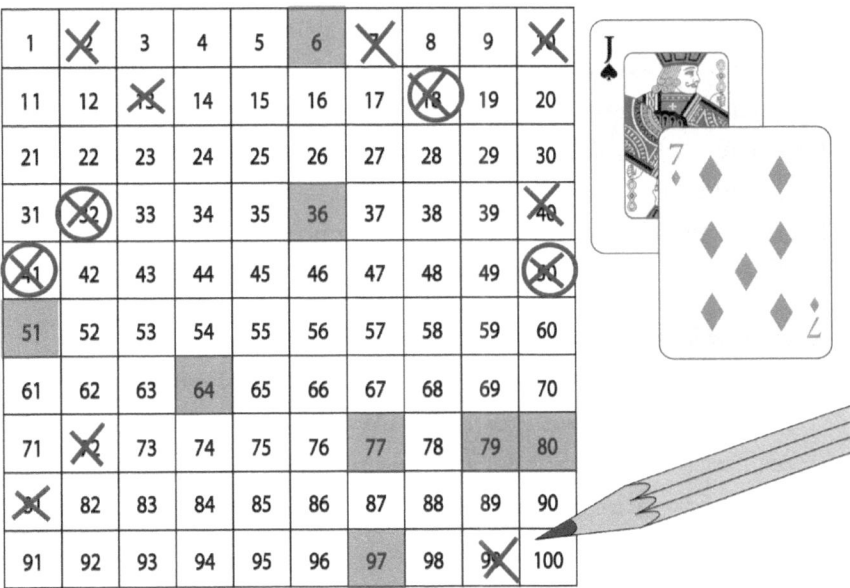

Towers give the Nazi guards a head start, but crafty prisoners can work around them. Which card would you use for your tunnel, and where would you place a guard?

After your initial move, you may build a tunnel in any unclaimed square. If one of the players hasn't made it onto the board, you may mark a square on their edge to let them in.

If one of your cards has no matching number square, use it for the guard (sleeping on patrol) and the valid card for your tunnel. But if both cards have no match, you lose that turn. For instance, there are only four cubic numbers on the hundred chart. After those are all claimed, kings become "lose a turn" cards.

Your goal is to dig a tunnel of squares that lets all players escape. Squares must be attached at their sides; squares meeting only at a corner do not count as connected. If the Nazi patrol blocks your path, you lose the game. But if you complete a tunnel that connects all players' initial squares, you win.

History

In the original Great Escape game, players compete against each other. But in real life, more than six hundred prisoners worked together to construct three tunnels under Stalag Luft III. One tunnel made it through. Seventy-six men escaped before the guards discovered the tunnel, filled it in, and launched a massive manhunt.

Unfortunately, only three of the escapees made it all the way across Germany to safety. But the project was important, even for those left behind. RAF pilot Jack Lyon explains:

> "It did a lot for morale, particularly for those prisoners who'd been there for a long time. They felt they were able to contribute something, even if they weren't able to get out. They felt they could help in some way and trust me, in prison camps, morale is very important."

My Nearest Neighbor

MATH CONCEPTS: comparing fractions, equivalent fractions, benchmark numbers.
PLAYERS: two or more.
EQUIPMENT: playing cards (one deck for two or three players, a double deck for four or more players). Calculator optional.

How to Play

Deal five cards to each player. Set the rest of the deck face down as a draw pile. Each number card represents its face value, aces count as one, and face cards as twelve.

Play four or more rounds:

◆ Closest to 0

◆ Closest to 1

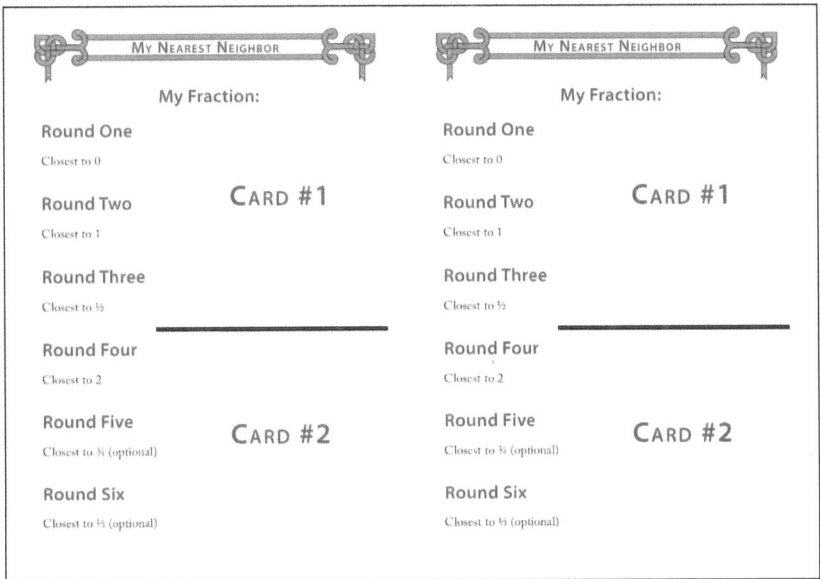

You don't need a gameboard, but you may find this reminder page from the printables file useful. Cut the page in half and give one sheet to each player.

- Closest to ½
- Closest to 2
- Closest to ¾ (optional)
- Closest to ⅓ (optional)

In each round, players lay down two cards from their hand to make a fraction as close as possible (but not equal) to the target number. Draw two cards to replenish your hand for the next round.

The player whose fraction is the nearest neighbor to the target collects all the cards played in that round. If there is a tie for closest fraction, the winners share the cards. In case of a rare three-or-more person tie, share as evenly as you can and leave any remaining cards on the table as a bonus for the winner of the next round.

After the last round, whoever has collected the most cards wins the game.

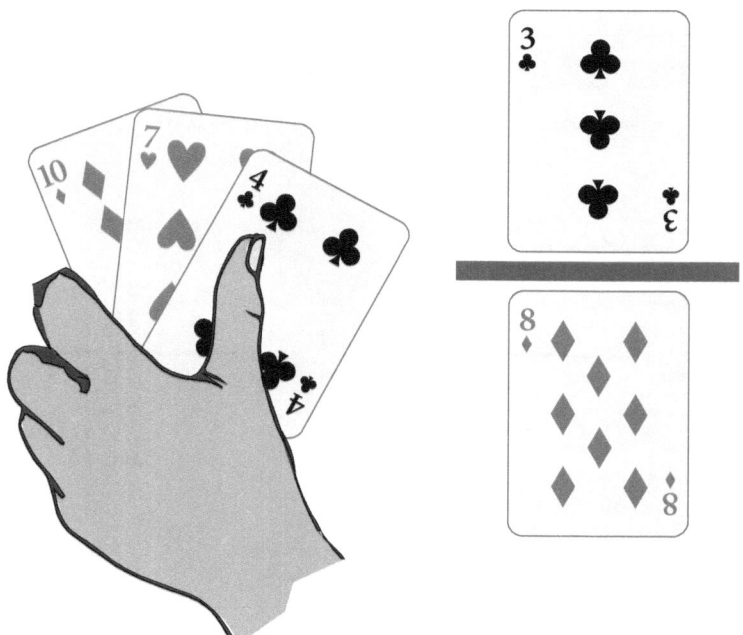

This fraction is close to ½, but the player could get closer. Can you see how?

History

Middle school teacher Christine Sullivan shared her favorite version of Fraction War at A Sea of Math blog, and I adapted it for this game. Math education hobbyist Joshua Greene of Three J's Learning blog suggested the current order for the target rounds, moving the hardest fractions to the end.

Words to Know

Common numbers and fractions that are relatively easy to work with are often called *benchmark numbers*. When they master the art of comparing fractions to benchmark numbers, our children develop strong mental math skills.

If you need help figuring out which fraction is the closest, you may enjoy the games and teaching tips in my earlier book *Multiplication & Fractions: Math Games for Tough Topics*.

Fractionator

MATH CONCEPTS: multiplying fractions, equivalent fractions.
PLAYERS: two players or two teams.
EQUIPMENT: printed gameboard, colored markers or a set of matching tokens for each player, two paperclips or other small tokens to mark the factors.

How to Play

The first player places a paperclip on any factor at the bottom of the board. The second player places the other clip on a factor—the same or different—and then marks the product of those two numbers by coloring the square or placing a token.

Fractionator

1/36	5/36	25/36	1/24	2/24	3/24
5/24	10/24	15/24	1/18	2/18	5/18
10/18	1/16	2/16	3/16	4/16	6/16
9/16	1/12	2/12	3/12	4/12	5/12
6/12	1/9	2/9	4/9	1/8	2/8
3/8	1/6	2/6	5/6	1/4	2/4
3/4	1/3	2/3	1/2	1	18/36

Factors: 1/6, 1/4, 1/3, 1/2, 2/4, 2/3, 3/4, 5/6, 1

Fractionator - Spiral

1	1/2	2/3	1/3	3/4	2/4
1/12	9/16	6/16	4/16	3/16	1/4
2/12	3/24	2/24	1/24	2/16	5/6
3/12	5/24	1/36	25/36	1/16	2/6
4/12	10/24	5/36	18/36	10/18	1/6
5/12	15/24	1/18	2/18	5/18	3/8
6/12	1/9	2/9	4/9	1/8	2/8

Factors: 1/6, 1/4, 1/3, 1/2, 2/4, 2/3, 3/4, 5/6, 1

My *Prealgebra & Geometry Printables* file includes traditional and spiral forms of the Fractionator and Decimator gameboards. By putting the more difficult math facts in the valuable center squares, the spiral version encourages players to stretch their skills.

You can place your token on the exact product of the numbers or on any equivalent fraction.

On each succeeding turn, a player shifts one paperclip to a new number and then marks the product of those two factors. If both players agree there are no possible moves, the player whose turn it is makes a fresh start by changing both clips.

Whichever player marks four (or more) squares in a row—horizontally, vertically, or diagonally—wins the game. The squares must touch each other at edges or corners, with no gaps. If neither player connects four, then the player who has the most sets of three in a row wins.

Variation

DECIMATOR: Practice multiplying and dividing decimal fractions by playing as above with one of the Decimator gameboards in the printable download file.

History

The answer when you multiply is called the *product* of the two numbers, so this game is also known as the Product Game.

Math professor and blogger John Golden calls the Product Game "the greatest math practice game ever." About the fraction version, he writes:

> "I did point out to the students that someone would tell them fraction division was hard, but they've already done it when they're figuring what to multiply ¾ by to get ⁶⁄₁₂."

Words to Know

Venn diagrams use overlapping shapes to create a visual representation of the relationships between different sets. Mathematician Georg Cantor defined a *set* as "a Many that allows itself to be thought of as a One."

While circles are the most commonly seen, a Venn diagram with four or more sets needs a more complex shape. For example, the Math Pickle website—a delicious source of mathematical activity ideas—has a five-set logo made from oblong pickle shapes.

What's My Rule?

MATH CONCEPTS: Venn diagrams, factors and multiples, divisibility, prime numbers, and other number properties.
PLAYERS: two or more.
EQUIPMENT: pencil and paper, or whiteboard and markers. Calculator optional.

How to Play

Choose one player to lead the game. The leader draws one or more large circles. With two or more circles, make them intersect in a Venn diagram. For each circle drawn, the leader must have a number-property rule in mind.

Each rule must be general enough to cover a reasonable range of guesses. For example, "numbers divisible by three" is a good rule because there are many common numbers that fit and also plenty that don't belong. But a rule like "multiples of 117" would rain shame on the leader's head and may result in banishment.

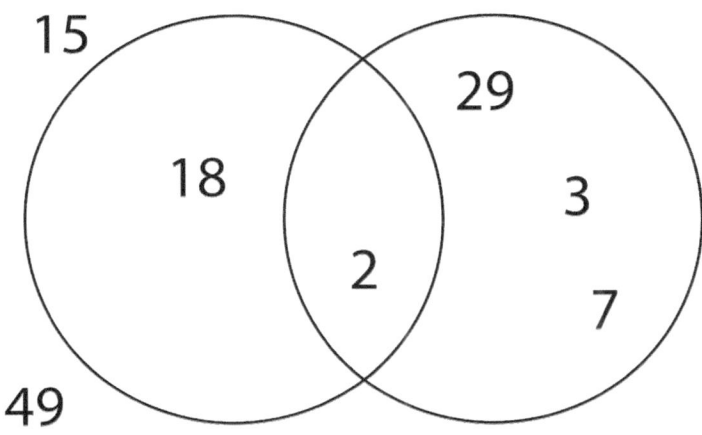

If numbers fit both of the rules, put them in the area where the circles intersect. Can you identify the two properties for this diagram?

NUMBER PROPERTIES ♦ 41

The other players take turns saying numbers. The leader writes each number in the appropriate circle of the Venn diagram. Numbers that fit more than one rule go in the region where those circles overlap. If the number doesn't fit any of the mystery rules, write it in the outer margin.

After the number is written down, the player who named it may try to guess the rule for one circle. If the guess is correct, the leader writes that rule next to the circle.

If a rule proves too difficult for players to guess, the leader may add number clues in that circle.

Variation

HOUSE RULE: Do you want to discourage wild guesses? Make a rule that the circle must contain at least three numbers before you allow players to guess its rule.

Number Riddles

MATH CONCEPTS: factors and multiples, divisibility, prime numbers, and other number properties.
PLAYERS: two or more.
EQUIPMENT: pencils and paper.

Set-Up

Agree on a suitable range of numbers for the game. Can players choose fractions? Negative numbers? Are any numbers too large or too small for consideration? Also, set a maximum limit on the number of clues allowed.

Players work separately to create their riddles. Choose a secret number that the other players will try to guess. Write four or more clues about your number.

If I chose the number fifteen, for example, I might write the following clues:

- ◆ My number is odd.
- ◆ The number has two digits.
- ◆ My number is less than fifty.
- ◆ It's a product of two prime numbers.
- ◆ The sum of the digits is six.

Arrange your number riddle clues in order from the hardest to the easiest. The hardest clue is the one that would apply to the most potential answers, like "My number is odd."

Every clue must eliminate at least one number. And there must be no other numbers that fit all your clues.

In testing my clues, I found two numbers that matched: fifteen and thirty-three. Oops! I need to add one more clue:

- ◆ You can make my number in change without using pennies.

How to Play

When all the players have finished creating their riddles, it's time to play.

One player at a time (the *riddler*) reads his or her clues aloud. The other players take turns trying to guess the secret number, one guess for each clue. The guessing team may discuss ideas before the player whose turn it is makes the guess.

You can play for the simple fun of guessing. Or keep score by counting how many clues you needed before you guessed correctly. With a group of players, track your score individually—but all the players of the guessing team get the same score for each riddle.

For example, imagine that Tom reads four clues before the team guesses his number. Everyone else adds four to their score. Tom adds nothing because he's the riddler. But when Janelle reads her clues, Tom will be on the guessing team, so he'll score points then.

After all the secret numbers are guessed, the player with the lowest total score wins the game.

Teaching Tip

The word-power potential of this game is virtually limitless. You may want to create a *word wall*—a poster or bulletin board on which you collect all the mathematical vocabulary words your child meets in each daily math lesson. Use these words to pose number riddle clues.

Digit Disguises

MATH CONCEPTS: addition, subtraction, multiplication, division, algebraic reasoning.
PLAYERS: two players or two teams. Best with teams.
EQUIPMENT: printed gameboards (optional), pencils and paper, or whiteboard and markers.

Set-Up

Each player or team needs a separate gameboard.

If you're not using printed gameboards, each player or team makes a chart with two columns labeled *Mine* and *Theirs*. List the letters A–J in each column. Next to each letter in your own column, write one of the whole numbers 0–9 in any mixed-up order. Fold your paper to keep that column hidden from your opponent. Slip a paperclip over the fold for greater security.

How to Play

On your turn, ask for the result of a basic arithmetic calculation involving two *different* letters. For example, you might ask:

$$A + B = ?$$
$$C - D = ?$$
$$E \times F = ?$$
$$G \div H = ?$$

But you may not ask for something like "I – I" or "J ÷ J."

The other player/team figures out the answer to your question, using their values for each letter, *but they do not tell you the number*. If the answer is on their chart, they tell you the letter for that value. Or they tell you that the result is not one of their letters.

THEIRS:

A =	X	1	2	3	4	5	6	7	8	X
B =	0	1	2	3	4	5	6	7	8	9
C =	0	1	2	3	4	5	6	7	8	9
D =	X	1	2	3	4	5	6	7	8	X
E =	0	1	2	3	4	5	6	7	8	9
F =	0	1	2	3	4	5	6	7	8	9
G =	X	X	X	3	4	5	6	7	8	9
H =	0	1	2	3	4	5	6	7	8	9
I =	0	1	2	3	4	5	6	7	8	9
J =	0	1	2	3	4	5	6	7	8	9

NOTES:

A + D = G

The gameboard in my printable file includes a chart of possibilities for the number code. Cross out each digit as you eliminate it. For example, if A + D = G, then none of those letters represent zero, and G cannot be smaller than three. Also, neither A nor D can be nine.

Players never say a number in response to a question. They only say a letter or "Not a letter."

Write notes to help you remember each clue. You'll need these to deduce your opponent's number code.

Then it is their turn to ask you for a clue about your secret code.

When one player/team claims to have broken the opponent's number code, the game is over. They read aloud the number values they have figured out. If they're entirely correct, they win the game. But if any of the answers are wrong, the opponent wins.

Variations

Don't end the game after the first code is guessed. Whether the guessing team won or lost, the other team keeps asking questions until they also break the code. If both teams fail, the game is a tie.

Words to Know

Arithmetic is the science of calculating with numbers. *Operations* are the things we can do with those numbers—such as addition, subtraction, multiplication, and division.

You can identify an arithmetic operation by naming its outcome:

- A + B = "The *sum* of A and B."
- C − D = "The *difference* between C and D," or to avoid confusion "the difference C minus D."
- E × F = "The *product* of E and F."
- G ÷ H = "The *quotient* of G and H," or to avoid confusion "the quotient G divided by H."

So why doesn't this game belong in the "Operations and Functions" chapter? Digit Disguises isn't about the operations themselves, but about deciphering the answers. To crack the secret code, players have to recognize the number properties hidden in each clue.

FOR BEGINNERS: If you'd like an easier game that uses similar reasoning, try What Two Numbers? in Chapter 5. Or you could allow players to give numbers in response to each question.

KEEPER OF THE CODE: (Mastermind-style) One player makes the code, and the other players try to crack it. Players take turns asking for clues, but they may discuss the answers and plan a strategy together. This variation is good for younger players, if you let them be the Keeper while older students or adults are guessing.

History

Math lecturer David Butler created this game and shared it on his blog Making Your Own Sense. He writes:

> *"I was struck by how quickly the logic got complicated, and then how quickly it all cascaded into finding everything when I finally got a few different numbers."*

Butler also shared a printable gameboard that can fold to stand on the table Battleship-style, hiding your secret code.[†]

[†] blogs.adelaide.edu.au/maths-learning/2019/09/21/digit-disguises

Integers

When we can tap into children's informal reasoning, we should do so because then we build on their understanding. They experience mathematics as a creative, sense-making activity.

But interestingly, we found that when we gave students contexts, such as owing money or increasing or decreasing elevation, they generally avoided using negative numbers.

I can talk about a debt as negative dollars or a loss of yards in football as negative yards, but when was the last time you watched a football game and someone said, 'Wow, that guy just gained negative three yards'?

This is not at all surprising, though. Historically, mathematicians resisted negatives for many centuries.

— Randolph Philipp and Jessica Bishop

Activity: Clothesline Math

MANY TEXTBOOKS MAKE A SERIOUS mistake in teaching about negative numbers: They present the rules and expect students to follow them. Of course, the books give explanations and justifications for each rule, but only a few children pay attention to those. Most simply memorize the rules, because to understand requires mental effort.

The whole approach is flawed. When we give students a rule, we give them permission not to think. All they need to do is remember our instructions.

But it is only by thinking—by struggling their way through mental difficulties—that our students can build a foundation of mathematical knowledge strong enough to support future learning.

That's fine, you may say. We'll supplement our math textbook with hands-on manipulatives and talk about real-life applications. Playing with blocks or colored chips will teach our kids to work with negative numbers, right?

Unfortunately, it's not that simple.

Studies indicate these don't really help. The concepts we *can* explain with manipulatives are just as easy to understand without them. And the things that make math so difficult—like, how in the world can negative times negative equal a positive number?—still cause trouble.

On the other hand, children do enjoy puzzles, even abstract ones like Sudoku. We can tap into our children's informal reasoning if we present integer math as a puzzle. We can give our kids freedom to test their own hypotheses. As they deduce the rules, students build a sturdy, reliable foundation of understanding.

We Have a Problem

Here's the challenge: Math should make sense. Numbers behave in predictable ways, most of the time, and deep in our hearts we believe they always should.

And with addition, that seems to be true. We can add two numbers and find an answer. Then we can take that number and add some more. We never run out of answers. There's always a bigger number.

But subtraction stymies us. Sure, there are plenty of subtraction problems we can solve, but even way back in second grade our teachers warned us of danger: "You can't subtract six from four."

Why not?

Isn't math supposed to work *all* the time?

How might we fix math so that subtraction will always make sense?

Children have seen thermometers with negative temperatures and timelines extending back into the past. And they've worked with number lines throughout elementary math. Students naturally propose a solution: Let the number line extend past zero into negative numbers.

Leave negative numbers as an abstract puzzle. Talk about the patterns your children see. Fold your paper at zero and notice the symmetry.

Give your kids time to play as they grow comfortable with the math.

WORDS TO KNOW: You can say the number −5 as "negative five" or "minus five," your choice. Both are acceptable. According to Keith Devlin, director of the Stanford Mathematics Outreach Project, mathematicians tend to prefer "minus." After all, an algebraic expression like "$-x$" could actually represent a *positive* number, if x itself is negative.

A Hands-On Number Line

To play with clothesline math, stretch a length of rope between two chairs or two hooks on a wall. Provide a basket full of clothespins. Give your students a stack of blank index cards or card-sized pieces of paper and some colorful pens or markers.

Write the number 0 on a card. Where shall we hang that on our clothesline? In math and science, zero is an arbitrary position. We can put it anywhere we like. Wherever we put zero, that becomes the "hinge" where we pivot between positive and negative.

How about a card for the number 1? Where should that go? Again, the position of one is arbitrary. It can go to the left of zero or to the right, as close or as far as we want.

But as soon as we put the 1 card in its place, that sets the *scale* of our number line. The position of every other number is fixed. 2 must be twice as far from 0, in the same direction as 1. −1 must be the same distance but in the opposite direction. And so on.

Have your students write several *integers* (positive and negative whole numbers) to hang on the line. How do they know where to place each number? They may need to reposition the numbers that were already placed (redefining the scale) to make room for their new numbers.

For a deeper level of thinking, try placing numbers on the line *without* defining zero or one. Any two numbers can set the scale for your number line. If you place a card for 2 and another for −5, can your students figure out where the −2 card should go? How do they know?

When students first think about negative numbers, they usually see the minus sign as an adjective modifying the number. But we can also look at the minus sign as a *unary operator*—that is, a mathematical verb that acts on a single input number. In contrast, addition and subtraction are *binary* operators because they only make sense with two input numbers.

If the minus sign is a verb, what action does it represent?

Create a series of cards with unary minus signs. Where should these hang on the number line?

$$3$$

$$-3$$

$$-(-3)$$

$$-[-(-3)]$$

etc.

Consecutive Capture Game

You need your number clothesline and one deck of regular playing cards.

Agree on which color represents negative numbers. At our house, we play with accounting colors: Black cards are positive numbers, and red cards count as negative. But some families like to use the electrician's standard that red is positive, black negative.

In this game, an ace may be used as one or twelve, player's choice. The number cards represent their face value. Count the jack as eleven, queen as zero, and king as a wild card that can be any number you wish.

Deal three cards to each player. Set the rest of the deck on the table as a draw pile.

On your turn, hang one of your cards in its proper position on the clothesline. If two cards have the same value, clip them together.

If your card makes a series of three or more consecutive integers, you capture all the consecutive cards. Remove them from the clothesline. But if there are two cards with the same value, only take one of those. Add the captured cards to your score pile.

Finally, draw a new card to replenish your hand.

When the deck is gone, turns continue until all cards are played. Whoever collects the most cards wins.

Place the cards in numerical order, like a number line with gaps.
If your card fills a gap, you capture all the consecutive cards.

Numbers Between the Integers

Stretch your children's ability to visualize number relationships. Bring out the blank cards and make new numbers to hang. Where would fractions, decimals, and mixed numbers (which combine a whole number with a fraction) fit on the line? Remember, any of these can be either positive or negative.

Do not accept "I don't know" for an answer. The point of clothesline math is for students to reason their way through the puzzle and wrestle it into submission. Ask your children to compare a troublesome number to other numbers they might find nearby.

Remind your students of what they've already noticed: The number line reflects itself across the zero point as in a mirror. In many ways, negative numbers behave the opposite of positive numbers. My daughter called them "backwards math."

Some children may find it helpful to remember the cartoon criminal Vector from the movie *Despicable Me*. Vector boasts, "I commit crimes with both direction and magnitude!"

What do *direction* and *magnitude* mean? How do these words relate to positive and negative numbers? An integer's magnitude is the distance from zero (the *absolute value*), and its direction is the positive or negative sign.

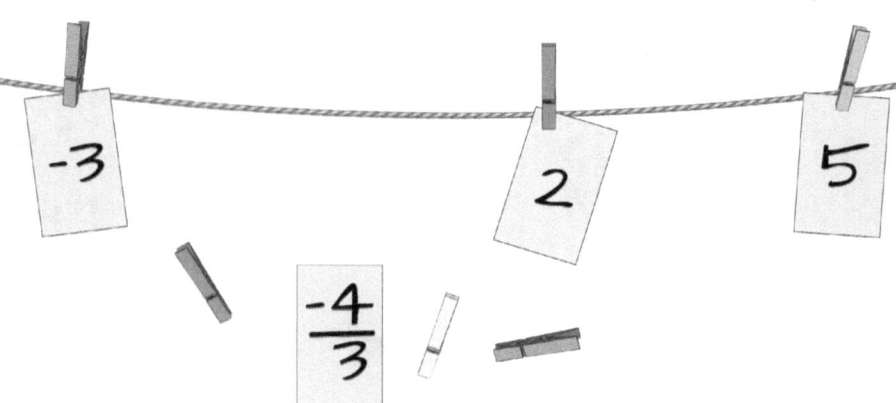

Based on the numbers already hung, where might be a reasonable position for this fraction card?

Conversation makes learning stick, as the students wrangle their fuzzy thoughts into solid words. For children (and adults) who have always considered math to be memorized rules, the experience of mathematical thinking—though difficult at first—can be as refreshing as a hike in pure mountain air.

As in hiking, so in math: It's not the destination that matters, but the journey. Yes, you could give your children a rule that would help them get correct answers, but that's no better than riding a helicopter up the mountain. Slow down and take the time necessary to let your children fully explore these concepts.

Addition Makes Sense

Math needs to make sense. How do these new numbers behave? What happens when we try to work with them?

Create cards with simple calculations and challenge your students to find positions for their values. Start with the relatively easy concepts of addition and subtraction.

- ♦ Where does "2 + 2" go on your number line?
- ♦ What about "−2 + (−2)"? Why?
- ♦ How about "4 + (−6)"?

Have children make up their own calculation cards to hang on the line. Talk about their thoughts. What do they notice about the positive and negative numbers? How do they figure out where each card should go?

Subtraction is the *inverse* (or mirror image) of addition. To subtract is like "un-adding" the amount. Can your kids make sense of that?

- ♦ Where does "7 − 4" go on your number line? How is this like un-adding four?
- ♦ Where does "7 − 7" belong?
- ♦ What about "−7 − (−7)"? Why? Does the rule that any number minus itself is zero work for negatives, too?

- Can they find a logical value for "4 − (−7)"?

WORDS TO KNOW: The *sum* is the answer when you add two numbers. The *difference* is the answer when you subtract. "Difference" means that subtraction tells you how different two numbers are—how far apart they are on the number line.

Integer Algebra

Let's try one more addition/subtraction challenge: missing-number equations. You can write these with blanks or use letters in place of the missing numbers. Either way, the puzzles nudge students toward algebraic thinking.

Can your students hang these cards where the missing number belongs on the number line?

$$3 = 5 + \_\_\_$$
$$-2 + \_\_\_ = -8$$
$$-4 = 1 - k$$
$$-7 - q = 9$$

Children love trying to stump adults. Let your students create missing-number cards for you to place on the number line. Think out loud as you solve each puzzle.

Multiplication Makes Sense, Too

Most adults remember struggling with the multiplication and division of integers. The rules felt so arbitrary. How can we help our children understand these topics?

Begin with the relatively easy calculations: positive times positive number, and positive times negative.

- Where does a card for "1 × 2" go on the number line? How about "2 × 2"?

- Where would you put "2 × (−2)"? Two −2s should be twice as far from zero as one −2 would be.

- How would your children think about "−3 × 5"? Remember that multiplication works in any order, so this is the same as "5 × (−3)."
- Can you figure out division? "12 ÷ 3" is located one-third of the way from zero to twelve. So "−12 ÷ 3" should be one-third of the way from zero to −12. Where does it go on your number line?

As subtraction is the inverse of addition, so division is the inverse of multiplication. Can your children explain how dividing by a number is like "un-multiplying"?

For example, −20 ÷ 4 means that we have a mystery number. When that mystery number is multiplied by 4, the answer is −20. So to solve the mystery, we need to un-multiply:

$$-20 \div 4 = ?$$
means that $-20 = 4 \times ?$

Again, have students make up their own calculation cards and missing-number puzzles to hang on the line. Discuss their thoughts. How did they decide where each card should go?

The End of the Game

Finally, pose the trickiest conundrum, the most difficult puzzle of all. If we want to trust mathematics, we need the ability to work with all the numbers in any combination. That means eventually we'll have to multiply (or divide) two negative numbers.

How can we ever wrap our minds around that?

James Tanton, founder of the Global Math Project, describes many people's experience with multiplying negative numbers:

> "In grade five I asked my mathematics teacher, 'Why is negative times negative positive?' The response I received was unenlightening to say the least: 'It just is!' This made me angry."

Check out Tanton's YouTube lesson "Why is negative times negative positive?" for the true mathematical answer to this question. Watch it with your kids, and enjoy the feeling of math making sense:

- Part I: youtu.be/nJqyr8h22nY
- Part II: youtu.be/eV6iYvd4KS0

The mathematical reason is this: It's true because we choose it so.

We *choose* to believe that numbers are consistent. And the *only* way numbers can be consistent is for negative times negative to come out positive. So our choice to believe in consistency forces that logical conclusion.

If your children have been using their heads all along, reasoning about the meaning of the numbers to understand each calculation, they'll be ready to understand this puzzle, too. They've noticed patterns in the way numbers behave, and those patterns will help them make sense of new situations.

Incidentally, this isn't really the end of your clothesline math game—or at least, it doesn't have to be. Clothesline math can prompt student thinking as you work your way through a variety of topics, even ahead into algebra. For more activity ideas, visit Chris Shore's Clothesline Math website.[†]

[†] *clotheslinemath.com*

Hit Me

MATH CONCEPTS: integer addition, absolute value.
PLAYERS: two or more.
EQUIPMENT: playing cards (two decks may be needed for a large group).

How to Play

Agree on which color represents negative numbers. Aces count as one, and face cards as ten. Choose one player as dealer.

The dealer gives each player one card face down and then turns one card face up beside each face-down card. *Players do not pick up their cards.* You may peek at your own face-down card as often as you like, but keep it hidden from the other players until the end of the round. The face-up card remains visible to all players.

Mentally add the numbers on your cards, taking into account both positive and negative integers. Your sum may go below zero.

When all players have had time to check their cards, the dealer asks each in turn whether they want a *hit*—an extra card dealt face up so everyone can see it. If you want the extra card, say "Hit me!" Add your new card to your running total, but don't say your sum out loud.

Last of all, the dealer may take a hit.

Then each player can ask for a second hit, and then a third, up to five hits (for a maximum of seven cards). Players may *hold*—stick with the cards they have—at any time, but they may not change their minds later and ask for a hit. The round ends when all players have either held or taken a total of seven cards.

At the end of the round, players turn their hidden cards face up and announce their scores.

The player with the lowest *absolute value*—the sum closest to zero, whether positive or negative—wins the round and becomes the new dealer. In case of a tie, the dealer hands the deck to any player who hasn't dealt recently.

Words to Know

Integers are the positive and negative whole numbers, plus zero. Zero is neither positive nor negative, but a neutral point between the two. *Absolute value* is always positive, because it measures only the *magnitude* (distance from zero), not the direction.

Variations

Do you hate relying on luck? Add a bit of strategy to the game by allowing the ace to count as one or eleven, player's choice.

The dealer could allow each player to take his or her hits all at once, then move to the next player. But with my students, that system allowed children too much idle time between turns.

History

My son's all-time favorite math game, Hit Me is a variation on the traditional gambling game Blackjack, in which players aim for twenty-one points. I originally called the game Zero, but my kids refused to acknowledge such a boring name.

I am often surprised at the score it takes to win a hand of Hit Me. If I have a sum of three or more, I almost always lose, unless I take another card. If I take the maximum number of hits, however, that is a sign of desperation. I remember one game when all the red cards came my way, for a total score of −37, as I kept trying without luck for at least one black number—and my math students laughed and cheered at every hit I took.

Strike It Out

MATH CONCEPTS: integer addition and subtraction.
PLAYERS: only two.
EQUIPMENT: printed gameboard or blank paper, pencils or markers.

How to Play

Draw a number line and label the numbers from −10 to +10. The first player strikes two numbers (draws a line through them) and circles their sum, saying the equation.

Then the second player must strike out the number the first player circled and one unused number, and then circle either their sum or their difference—which also must be an unused number. Be sure to say your equation out loud, so the other player can check your mental math.

Once a number is struck out by either player, it can't be used again.

The first player must add, but after the first turn you may add or subtract. Continue to take turns striking two numbers (one of which must be the answer from the previous turn) and circling their sum or difference.

Eventually the numbers run out or the remaining numbers are

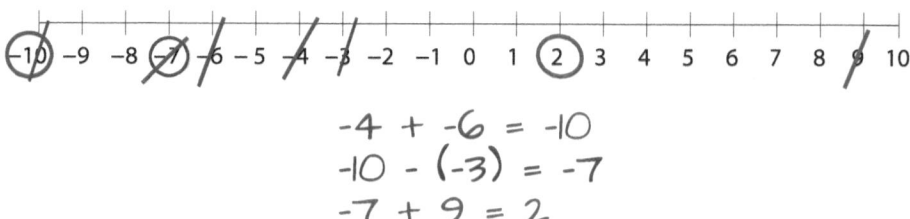

A sample game of Strike It Out. I've written equations to show each turn's play, but players may explain their moves orally.

impossible to reach. Whoever makes the last legal move wins the game.

History

Strike It Out is from the Nrich Maths website, which is full of mathematical activities and resources for teachers and school-age students. At age seven, John Domoradzki (son of author Amber Domoradzki) suggested extending the game into negative numbers.

Integer Solitaire

MATH CONCEPTS: integer addition and subtraction.
PLAYERS: one or more (a cooperative game).
EQUIPMENT: playing cards, large sheet of poster board (optional).

Set-Up

Draw the four equations below on a sheet of poster board, large enough that playing cards fit in the blanks.

If you don't have poster board, draw the equations on paper (or use the gameboard from my *Prealgebra & Geometry Printables* download file) so players can write in their numbers. You may want to slip the gameboard into a clear page protector or laminate the paper for use with dry-erase markers, to make it easier to move the numbers around.

Agree on which color represents negative numbers. Aces count as one, number cards at their face value. The jack, queen, and king are eleven, twelve, and thirteen, respectively.

How to Play

Deal out eighteen cards and set the rest of the deck aside. Arrange these cards in the boxes on your gameboard. There are only fourteen blanks, so you won't use all the cards.

Can you make four true equations? If so, you win.

Variation

If you succeed with eighteen cards, try the game again with seventeen. Can you do it with sixteen cards? Is fifteen enough to make it work?

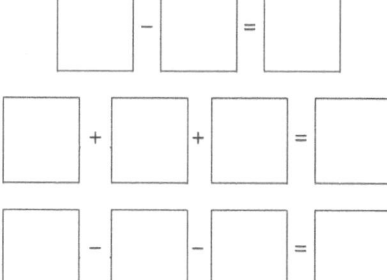

History

Math instructor Kent Haines, who created this puzzle game, writes:

> *"I love this game because it doesn't require a lot of supplies, can be played in fifteen minutes, and remains challenging even after a student has mastered integer addition and subtraction.*
>
> *"I picked the starting amount of cards on intuition. I have played this game for five years with dozens of students, and I have yet to see a combination of eighteen cards that is unsolvable."*

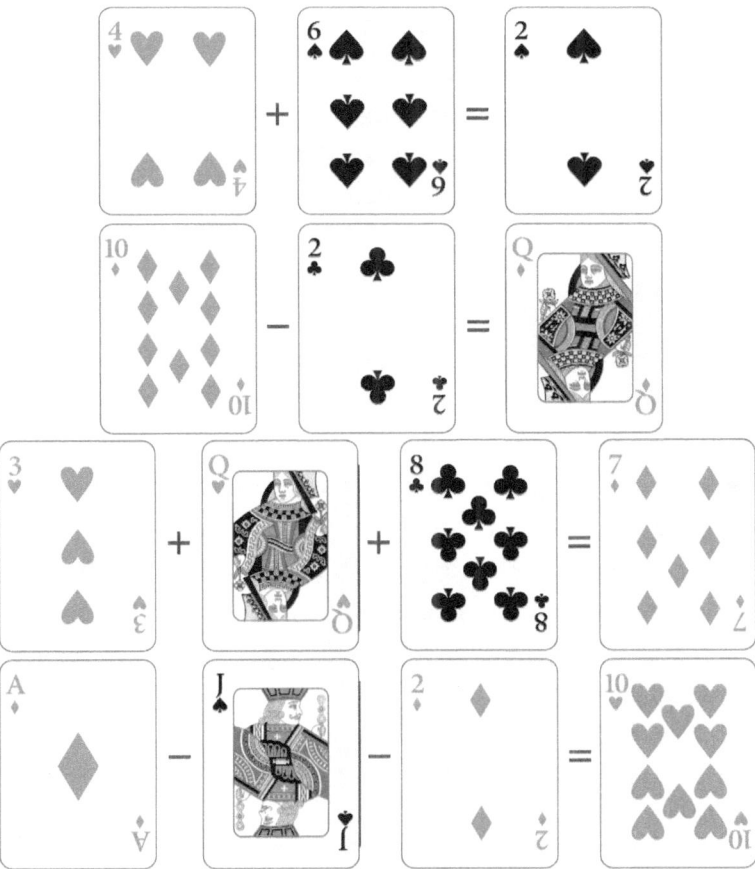

The four equations of Integer Solitaire,
with a winning arrangement of cards.

Fight for the Center

MATH CONCEPTS: data sets, mean, median, mode, range.
PLAYERS: two or more players or teams.
EQUIPMENT: one homemade scoreboard per player/team, two decks of playing cards, pencils and scratch paper for calculations. Calculator optional.

Set-Up

Each player or team creates a scoreboard by folding a sheet of paper into quarters. Unfold the paper and write one word on each section: Mean, Median, Mode, and Range.

Remove the jacks and kings from your playing card decks, but keep the queens to represent zero. Aces count as one, number cards at their face value. Agree on which color represents negative numbers.

How to Play

Shuffle the card decks together and deal four goal cards to each player or team. Players look at these cards and then place them face down on the table beside their scoreboards. You may peek at your own goal cards whenever you wish, but keep them secret from other players until you are ready to score.

Turn six cards face up on the table as the initial data set and arrange them in order from least to greatest. All players share these cards.

Then deal a hand of five cards to each player or team. Place the rest of the deck face down for drawing cards. As you play, a face-up discard pile will grow next to the draw pile.

On your turn, you have two choices:

- ♦ Draw a card. Discard one number from the data set and replace it with a card from your hand, rearranging the set if necessary to keep the numbers in sequence. If one of your

In this data set, the mode and median are both +2, the range is 8, and the mean is ⅔.

goal cards matches the new mean, mode, median, or range of the data set, turn that card face up and place it in the proper section on your scoreboard.

♦ Or switch out one of your secret goal cards. Put the goal card in your hand, and replace it with a different card from your hand. You do not draw or discard, and you may not claim a goal on this turn.

At the end of your turn, you should again have five cards in your hand.

You can only score each of the data measures once per game. If you place a card on the median, for example, your other goal cards have to fit the mode, mean, or range.

The range is always a positive number, so you may take the absolute value of your goal card. For the other measures, your card must match exactly.

You may only claim the mode if there are at least two copies of that number in the data set.

If the draw pile runs low, shuffle the discards back into the deck to keep playing. The first player or team to score three of their four data measures wins the game.

Variations

HOUSE RULE: Do your children find it too hard to hit their goals? Give them more flexibility by allowing players to change the number of data

Words to Know

Statistics is the art of collecting, measuring, and interpreting data. A *data set* can be any collection of numbers or other information.

We can describe a numerical data set using several *measures of central tendency,* sometimes called "averages." The most common of these are:

- *Mean* (or *arithmetic mean*) = the value it would be if all the data were collected together and then shared out evenly. To calculate the mean, find the sum of the data values and divide by the number of data points.
- *Median* = the middle value when the data is listed sequentially from least to greatest. If you have an even number of data points, take the mean of the two middle values.
- *Mode* = the most frequently repeated data value. A set of data may have more than one mode.

Beware pundits or marketers who try to mislead you by citing the "average" of some data set without telling you *which* measure of central tendency they are using. And while the arithmetic mean is the most commonly used average, it may not be the most helpful in understanding a particular set of data.

One more way to describe a data set is the *range,* which is the distance the data would cover on a number line. To calculate the range, subtract the smallest data value from the greatest value.

points. Instead of replacing a card, they may add a card from their hand to increase the size of the data set.

Or they don't have to draw and play from their hand. They can just remove a card to the discard pile, making the data set smaller. This counts as a regular turn, so they may score a goal.

If you play this variation, you'll probably want to set limits such as a maximum of ten and a minimum of four data cards.

FLASH GAME: The first player to score any data measure wins. Or agree in advance on one data measure players need to match.

History

High school math teacher James Cleveland created this game and shared it on his blog, The Roots of the Equation.

Words to Know

The *area model* for multiplication is a visual thinking tool. We can represent the product of any two numbers as the area of a rectangle with sides the length of those numbers.

It's easy to show why this works with positive whole numbers by coloring squares on square grid paper. One number tells how many rows to color, and the other number tells how many squares to include in each row. The rectangle's area, the total number of squares, is the product of the two numbers.

The area model works with fractions, it's great for understanding algebraic multiplication, and it's a key idea in calculus. Best of all, it works for negative numbers, too, if you're willing to accept the idea of a negative area. This is how James Tanton makes sense of the rules for multiplying integers, as mentioned in the Clothesline Math activity.

Grid Fight

MATH CONCEPTS: integer multiplication, area model for multiplication.
PLAYERS: two players or two teams.
EQUIPMENT: printed gameboard or square grid paper, playing cards, pencils or colored markers.

Set-Up

Print a gameboard for the players to share. Or make your own gameboard on square grid paper: two 12 × 12 squares separated by a "zero" line.

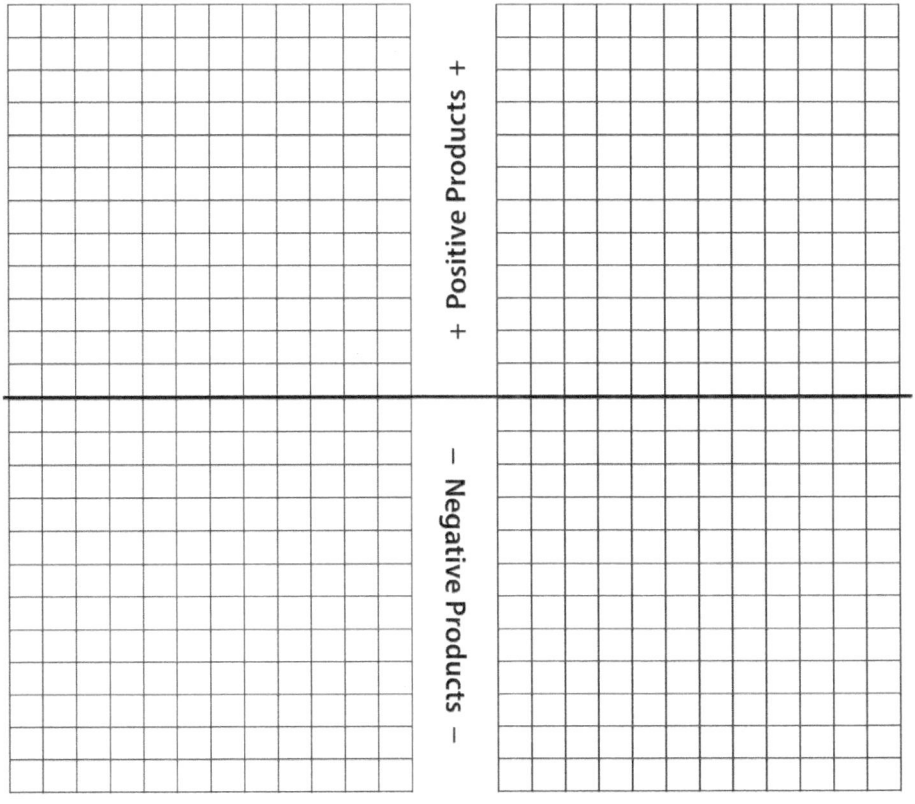

The Grid Fight page from my printable file has room to play two games.

Label one square positive (+) and the other negative (−). Choose which player or team will use the positive side of the gameboard grid and who takes the negatives.

Remove the face cards from your deck. Agree on which color represents negative numbers.

How to Play

Deal three cards to each player. Place the rest of the deck face down as a draw pile.

On each turn, both players make their move at the same time. Choose one card to play and hold it face down on the table in front of you. When both players are ready, turn up the chosen cards and multiply the numbers.

If the product is greater than zero, the positive player colors a rectangle of that size on the positive side of the grid. If the product is negative, that player colors a rectangle on the negative side.

Your goal is to fill in as many complete rows as you can, as in the game Tetris. But if you can't fit your rectangle on the grid, you lose that turn.

Both players discard their used cards and draw to replenish their hands. Play twelve turns. Whoever fills in the most rows wins the game.

Variation

Do you want a longer game? Play on both halves of the gameboard at once. Players may color their rectangle in either square (on their own side of the zero line) wherever they can fit it in. The game continues until the grids are so full that players lose three turns in a row.

History

I owe this game (and many others!) to the amazing John Golden. Visit his Math Hombre Games webpage for more ideas.[†]

[†] *mathhombre.blogspot.com/p/games.html*

Honeycomb

MATH CONCEPTS: integer addition, integer multiplication.
PLAYERS: two players or two teams.
EQUIPMENT: printed gameboard or hexagon grid paper, deck of playing cards, pencils or markers.

Set-Up

Print a gameboard or a page of hexagons for the players to share. The gameboard has room for four short games. A full-page grid allows one longer game.

Decide which player will take negative numbers, and the other takes positive.

Also decide which color card represents negative numbers. Remove the face cards and all numbers greater than six. Spread the remaining cards face down to make a fishing pond.

How to Play

The negative player goes first. Draw a card from the fishing pond. With the resulting positive or negative number, you can either:

♦ Add that value by writing it into any open hexagon.

♦ Multiply an existing hexagon number by that value to create a straight line of adjacent hexagons.

You multiply by *replicating* the existing number, copying it into open spaces connected to the original. The line must include as many hexagons as the value you are multiplying by. And if your number is negative, change the sign of the numbers in the row. You can multiply only if there are enough open spaces to make a straight line. Multiplication rows may not turn.

Suppose you drew −4. You might…

INTEGERS ♦ 73

Add -4 anywhere

Or multiply the 5 to make a row

-5
-5
-5
-5
-4
-1

You have two choices: add or multiply.
Multiply a number by making copies in a row.
If you are multiplying by a negative number, remember
to change the sign. "−4 × 5" is the same as four −5s.

- Add a –4 in any unclaimed hexagon. Or…
- Notice that one hexagon has a 5 with plenty of blank spaces around it. Write three more 5s in a row and put minus signs on them all, for a total of four –5s. Or…
- Notice a space with –1 and multiply it into a row of four +1s.

After your turn, place your card back in the fishing pond and mix thoroughly.

Play until the gameboard is filled. If the sum of all the numbers on the gameboard is less than zero, the negative player wins. If the sum is greater than zero, the positive player wins.

The losing player gets to choose positive or negative for the next game.

Variations

If you're playing on a full-page hexagon grid, you may wish to include all the number cards in your fishing pond. Remove the face cards, or count them as wild cards.

House Rule: Do you wish multiplication lines could turn to fit on the gameboard? Let multiplication run like a curvy snake, as long as all the replicated copies connect to the original number with no gaps.

History

Honeycomb was created by Nick Smith and John Golden.

Integer War

MATH CONCEPTS: integer addition, integer multiplication, fractions, powers (exponents), absolute value.
PLAYERS: two to four.
EQUIPMENT: playing cards (one deck per player). Calculator optional.

How to Play

Players agree on which color represents negative numbers and which Integer War version they wish to play. Aces count as one, number cards at their face value. The jack, queen, and king count as eleven, twelve, and thirteen, respectively.

Players each stack a deck of cards face down on the table or floor. Then all turn their top two cards face up and mentally perform the agreed-upon calculation, announcing their answers out loud. The winner captures all the cards showing. Each player keeps a pile of prisoner cards.

If there is a tie for greatest absolute value, push all the cards into a center pile. Then the players go on to the next skirmish, again turning up two cards from their decks. The winner of that skirmish also takes the center pile. If there is no winner this time, repeat until someone wins a battle and captures the whole mess.

When the players have fought their way through the entire deck, count the prisoners (or compare the height of the stacks). Whoever has captured the most cards wins the game. Or shuffle the prisoner piles and keep playing until one player captures all the cards, or until all the other players concede.

Variations

The classic children's game of War adapts easily for practicing a variety of integer operations. To provide a greater challenge, remove the aces, twos, and tens. This gives each player a deck full of the toughest problems.

With some variations, players will have only one card remaining on the last turn. Turn up these cards, and players race to make the greatest (or least) absolute value using all the numbers showing and any math manipulation they can do mentally. If two or more players make the same value, the first one to say it aloud wins the skirmish.

INTEGER ADDITION/SUBTRACTION WAR: The greatest sum (or difference) wins the skirmish.

INTEGER MULTIPLICATION WAR: The greatest product wins. Remember that two negative numbers make a positive product.

FRACTION WAR: Use two cards to make a fraction with the smaller card as the numerator. (This is the same as dividing one card by the other.) The winner is the fraction with the greatest absolute value—that is, the farthest away from zero.

POWER WAR: Use one card as the base and the other as an exponent, and the greatest value wins. Let students use a calculator as needed.

MATH WAR TRUMPS: Players draw three cards and alternate choosing trumps for the math card battles. Instead of flipping the cards face up, pick them up and look at them in your hand. The player whose turn it is says which operation to do and whether the high or low value wins. Then all players choose two of their cards for the battle, lay those on the table, and draw to replenish their hands.

WILD WAR: Players turn up three cards and may do whatever math manipulation they like with the numbers. But no calculators allowed, so you cannot use anything too complex to work in your head. The player with the greatest absolute value (farthest away from zero) wins.

REVERSE WILD WAR: Players turn up three cards and may do whatever math manipulation they like with the numbers. The answer with the *least* absolute value (closest to zero) wins the skirmish.

History

I tire quickly of the Math War game because there's no strategy involved—except in the Trumps or Wild War variations—but the students in my math classes always enjoy playing. It's a quick and easy way to use up a few extra minutes at the end of a lesson.

Math War is the simplest variation of a whole family of traditional math-teacher games involving a pre-defined math operation. Players choose random numbers to fit into that operation, trying to get a particular value—usually the greatest, least, or closest to a target number. Teacher Jenna Laib coined the name Number Boxes for these games, and you can read more about them on her Embrace the Challenge blog.[†]

As your students move into high school, you can practice algebra, geometry, and trigonometry topics with Math War. Check the post on my Let's Play Math blog for a list of free teacher-created decks of special math cards.[‡]

[†] *jennalaib.wordpress.com/one-of-my-favorite-games-number-boxes*
[‡] *denisegaskins.com/2020/05/06/math-game-war-with-special-decks*

Four in a Row

MATH CONCEPTS: integer multiplication, factors, multiples, algebraic multiplication.
PLAYERS: two players or two teams.
EQUIPMENT: printed or homemade gameboard, colored markers or a set of matching tokens for each player, two paperclips or other small tokens to mark the factors.

Set-Up

Choose a Four in a Row gameboard for the players to share. Or have students work together to create their own gameboard:

- On a blank sheet of paper, draw a 6 × 6 array of squares.
- In the space below the array, each student writes four factors.

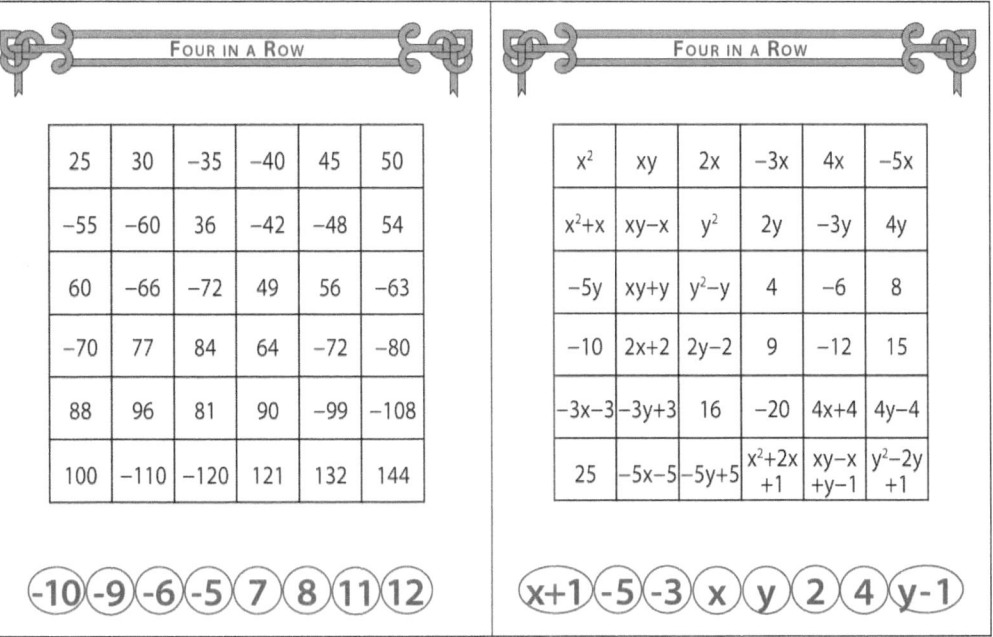

The free download file offers a variety of gameboards.
Choose the difficulty level that best fits your students.

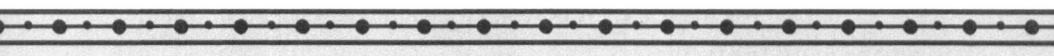

Words to Know

An *algebraic variable* is a letter or symbol that stands in for some other value.

Often we think of variables as having an unknown or yet-to-be-determined value, as in a "solve for *x*" algebra exercise. In other cases, a variable may have a known-but-changing value, as when NASA engineers calculate the speed and acceleration of a planetary probe.

Or the value of a variable may depend on the situation. Consider the variables for length and width in the formula "$A = l \times w$" for the area of a rectangle. These measurements are *fixed* (or *constant*) for any rectangle, but they change from one rectangle to the next.

Scientific formulas often include a combination of variables representing fixed values and values that can change.

Factors may be positive or negative and may include fractions, decimals, or even algebraic variables.

♦ Then take turns writing the product of any two factors into one gameboard square. A few duplicates won't ruin the game, but try to avoid writing a product that's already been used.

When all the squares are full, you're ready to play. You may want to slip your gameboard into a clear page protector or laminate it for playing with dry-erase markers.

How to Play

The first player places a paperclip on any factor at the bottom of the board. The second player places the other clip on a factor—the same or different—and then marks the product of those two numbers by coloring the square or placing a token.

On each succeeding turn, a player shifts one paperclip to a new number and then marks the product of those two factors. If both players agree there are no possible moves, the player whose turn it is makes a fresh start by changing both clips.

Whichever player marks four (or more) squares in a row—horizontally, vertically, or diagonally—wins the game. The squares must touch each other at edges or corners, with no gaps. If neither player connects four, then the player who has the most sets of three in a row wins.

Variation

PATHWAYS: One player "owns" the top and bottom of the gameboard, while the other player claims the right and left sides. The first player who can mark a path of squares across the board (top to bottom for one, side to side for the other) wins the game. The pathway squares must all connect by sharing a side or corner.

History

I first saw this type of four-in-a-row game in 1987 on the old *Square One TV* show. The basic integers gameboard comes from John Golden's "Integer Games" blog post. Bill Lombard and Brad Fulton published the challenge gameboards in *Simply Great Math Activities: Algebra Readiness*.

The Pathways variation—which is a cross with the strategy game Hex—comes from Marilyn Burns's Math Blog.

Operations and Functions

Earlier we considered the argument, 'Twice two must be four, because we cannot imagine it otherwise.' This argument brings out clearly the connexion between reason and imagination: reason is in fact neither more nor less than an experiment carried out in the imagination.

People often make mistakes when they reason about things they have never seen. Imagination does not always give us the correct answer. We can only argue correctly about things of which we have experience or which are reasonably like the things we know well.

If our reasoning leads us to an untrue conclusion, we must revise the picture in our minds and learn to imagine things as they are.

When we find ourselves unable to reason (as one often does when presented with, say, a problem in algebra), it is because our imagination is not touched. One can begin to reason only when a clear picture has been formed in the imagination.

— W. W. Sawyer

When children make their own math, that builds solid understanding. Emma Baldwin, a student from my local homeschool co-op, created this seesaw balance puzzle. How much does each playground ball weigh?

Activity: Toy Tug of War

Why do students struggle with algebra? According to master teacher W. W. Sawyer, the problem is all in our heads: "One can begin to reason only when a clear picture has been formed in the imagination."

Hands-on puzzles help children form clear mental pictures of algebraic concepts. For instance, "pan balance" equations have a long history in math textbooks and form the basis of Henry Borenson's Hands-On Equations curriculum. The mental image of a balance scale highlights the "this is the same amount as that" meaning of the equal sign.

We don't need a special curriculum to play with algebra. With only a few toys and bits of string, we can bring algebraic equations to life for our kids.

Gather two or three distinct classes of small toys such as miniature soldiers, Lego people, or plastic dinosaurs. If you don't have suitable toys, you could use small origami figures or cut simple geometric shapes like squares, triangles, and circles.

You'll also need a few pieces of string or thin wire. And collect scraps of plain paper for writing signs, or use a whiteboard with erasable markers.

Three Simple Steps

For this activity, we assume that all the members of a class pull with the same force. All the toy soldiers have the same strength, the dinosaurs are interchangeable, and so on.

To create a Toy Tug of War puzzle, follow these steps:

(1) Think of a secret number that represents the strength for each class of toy. This can vary from one puzzle to the next, and it doesn't have to match reality. For example, I might imagine the soldiers having strength value 3 while the dinosaurs each pull with 2 points of force.

(2) Arrange the toys on a piece of string so their tug of war is balanced, the same strength on each side. Mark this contest with an equal sign. In my example, four dinosaurs have 8 points of strength. So I can balance them with two soldiers and another dinosaur.

(3) On a second piece of string, arrange an imbalanced tug of war. Put a question-mark sign in the middle. For example, I might pit two soldiers against a dinosaur, as shown below.

Can your students tell who wins the second tug of war? How do they know?

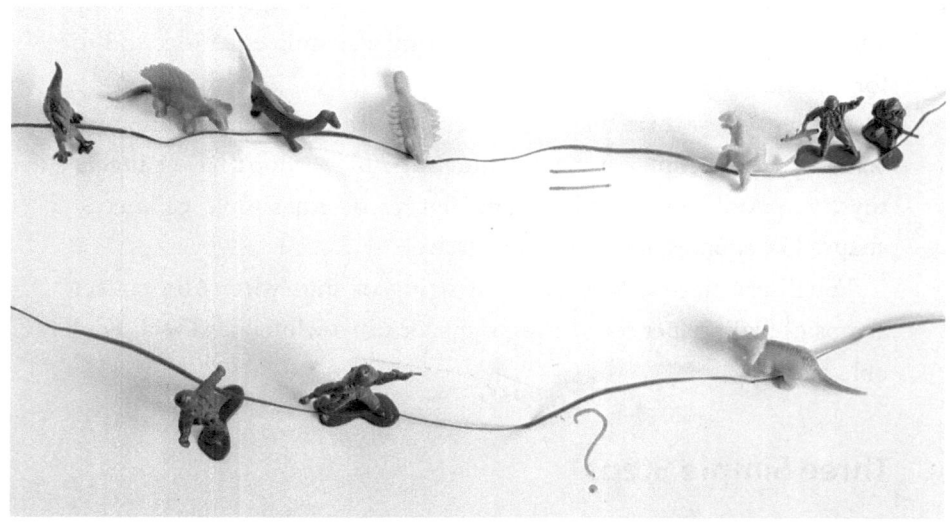

Dinosaurs versus soldiers. Who will win?

$$4 \text{ dinosaurs} = 1 \text{ dino} + 2 \text{ soldiers}$$
$$2 \text{ soldiers (?) } 1 \text{ dinosaur}$$

To solve this puzzle, notice that the dinosaur on the right-hand side of the first tug equation balances one dino on the left. If the tug is equal, then two soldiers must pull with the strength of three dinosaurs. So who wins the second tug of war? The dinosaur doesn't stand a chance.

Notice that we don't ask our students to find the strength values of our toys. Those numbers were only to help us create a puzzle that makes sense. My tug-of-war puzzle above would work with many values of toy-strength:

$$\text{soldiers} = 3, \text{dinosaurs} = 2$$
$$\text{soldiers} = 6, \text{dinosaurs} = 4$$
$$\text{soldiers} = 1, \text{dinosaurs} = ⅔, \text{etc.}$$

Puzzle Tip 1

Adding a new toy to both sides of a balanced tug-of-war equation doesn't affect the solution. It only makes the puzzle look harder.

In the scenario below, I thought of the red ninjas as being worth 1 point each, and the Lego people as 2 points. The lizards could have any value.

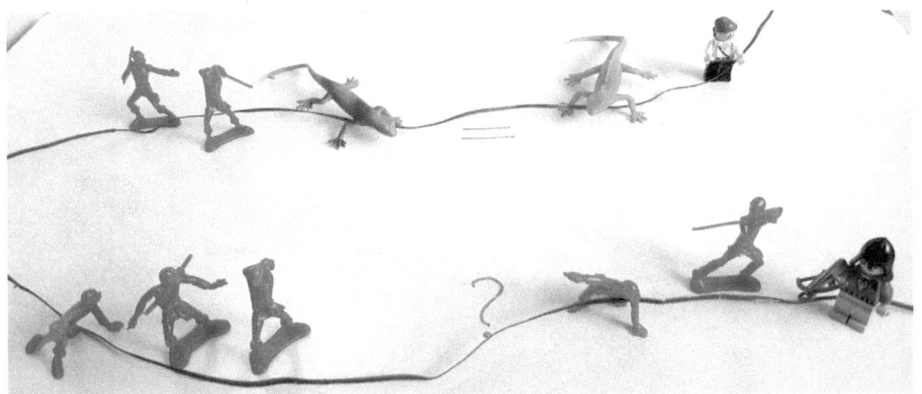

Red ninjas and Lego people. The lizards are distracters.

$$2 \text{ ninjas} + \text{lizard} = \text{lizard} + \text{Lego guy}$$
$$3 \text{ ninjas} (?) 2 \text{ ninjas} + \text{Lego guy}$$

Because the lizards balance each other, we can ignore them. Therefore, in this puzzle one Lego person pulls with the same strength as two ninjas.

So the right-hand team on the second tug of war must win.

Puzzle Tip 2

As the strength relationships get more complicated, you'll need to set up more than one balanced tug-of-war equation before posing a puzzle. With three classes of toy figures, you often need two balanced equations. Or with four classes of toys, you generally need three balanced equations to establish the strength relationships.

With imaginary strength values of 1, 3, and 5 points, I set up the following puzzle. Can you solve it?

Superheroes, dinosaurs, and black ninjas. Who will win?

$$2 \text{ superheroes} = \text{dinosaur} + \text{ninja}$$
$$\text{superhero} + 2 \text{ ninjas} = \text{dinosaur}$$
$$2 \text{ ninjas (?) superhero}$$

To solve this puzzle, start with the second tug-of-war equation. One dinosaur pulls with the same strength as the team on the left. So we can replace the dino in the first equation with a similar team. That transforms the first tug-of-war equation into:

2 superheroes = superhero + 3 ninjas

Which means one superhero must pull with the same power as three ninjas. So who wins the third tug of war?

Puzzle Tip 3

Thinking of multiplication can lead to more complex tug-of-war puzzles. In the next scenario, both sides of the first equation represent 21 points. Can you see how?

Lego people, soldiers, and giant lizards.
This one takes some thought.

3 Lego people = 7 soldiers
1 lizard = Lego guy + 2 soldiers
1 lizard + 3 soldiers (?) 3 Lego people

If three Lego people match seven soldiers, then we can think of both sides of the first tug-of-war equation as representing the multiplication relationship $3 \times 7 = 21$. The three Lego guys must be worth 7 points each, and the seven soldiers count as 3 points each.

We can use those values in the second tug-of-war equation to show that the lizard pulls with 13 points' worth of strength. Substituting these numbers into the last tug of war will show which side wins.

Or, we can solve the whole thing without using numbers. By splitting the seven soldiers in the first equation into three groups (plus an extra guy), we can see that each Lego person pulls with a bit more force than two soldiers.

Because of the second tug-of-war equation, we can imagine replacing the lizard in the last tug of war with a Lego person plus two soldiers. And also replacing the three Lego guys on the right with seven soldiers from the first tug equation. That would transform the puzzle into this:

$$\text{Lego guy} + 5 \text{ soldiers} \;(?)\; 7 \text{ soldiers}$$

Since the Lego person is stronger than two soldiers, the team on the left wins this tug of war.

After your children have worked a few of your tug-of-war puzzles, encourage them to create a puzzle for you to solve.

Teaching Tip

Did you notice how well using words in the equations worked to communicate what was happening in each problem?

The next time your children get stuck on a math story problem, encourage them to try writing equations with words. I call it *word algebra,* and it's one of my favorite thinking tools.

What Two Numbers?

MATH CONCEPTS: addition, multiplication, inverse operations, positive and negative numbers.
PLAYERS: two or more.
EQUIPMENT: no equipment needed.

How to Play

The leader chooses any two numbers and mentally figures their product and sum. Then the leader asks, "What two numbers multiply to make ___ and add up to ___?"

The leader may choose any two operations to ask. For example:

- What two numbers add up to 15 and multiply to make 50? (5 and 10)
- What two numbers have a difference of 2 and a sum of zero? (−1 and 1)
- What two numbers have a product of ⅙ and also have a difference of ⅙? (−⅓ and −½) or (⅓ and ½)

The other players race to find the numbers. The first player to name them correctly gets to lead the next round. Or with two players, just take turns trying to stump each other.

Remember to consider both positive and negative numbers when creating your puzzle. Or make it extra tricky with fractions or decimal numbers.

Words to Know

In case you need a reminder of the answer words: The *sum* is what you get when you add two numbers. The *difference* is when you subtract. You multiply to find the *product*, and divide to get the *quotient*.

History

An anonymous teacher left this game in a comment on my blog and added, "I love playing math games in the car with a 'captive audience.'"

The What Two Numbers? game was designed to prepare students for factoring quadratic equations, where one knows a sum and product and must deduce the original numbers. When we offer players a wider choice of operation clues, it becomes a more creative problem-solving game.

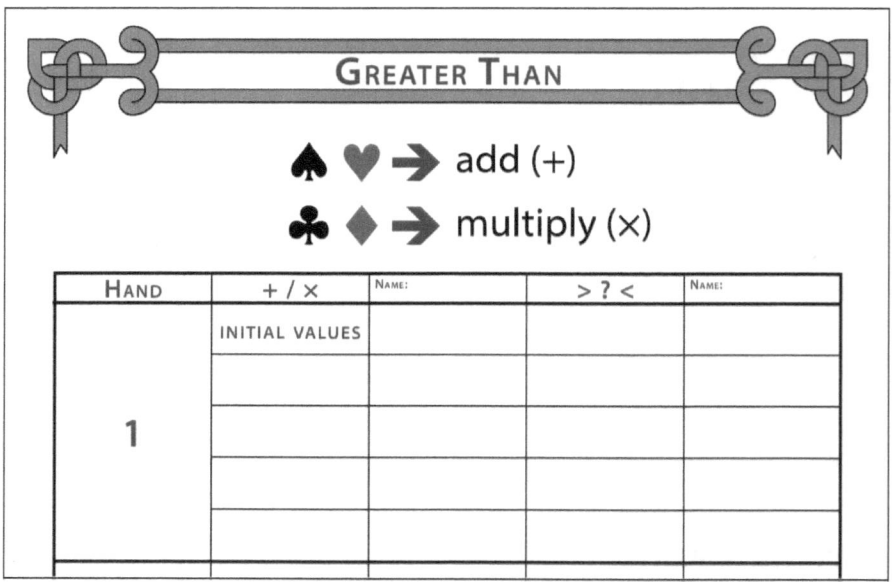

You can use plain paper to keep score in the Greater Than game, but you may like the gameboard from my *Prealgebra & Geometry Printables* file because it reminds you what each suit represents.

Greater Than

MATH CONCEPTS: integer addition, integer multiplication, working with inequalities.
PLAYERS: two players or two teams.
EQUIPMENT: printed gameboard or sheet of paper, one deck of playing cards, pencils or markers.

Set-Up

Agree on which color represents negative numbers. Face cards count as ten, aces as eleven.

The suits also represent operations:

♦ Spades and hearts = add the number.
♦ Clubs and diamonds = multiply by the number.

Print a gameboard for players to share. Or prepare a scoresheet with three columns: one for each player's running score, and a column in the middle for the inequality sign. Choose a player to deal the first hand.

How to Play

Deal four cards to each player or team. Players each choose one card as their initial value, holding the card face down in front of them. At the dealer's signal, both players reveal their cards.

If the two cards played have the same value, players return them to their hands and choose a different starter card.

Write the players' initial numbers in their columns. In the middle column, draw an inequality symbol (< or >) with its open end toward the greater value.

The non-dealer plays first. Choose one card from your hand and lay it on the table. Write the operation represented by that card in the first

Words to Know

An *expression* is a symbol or set of symbols that represent a numerical or algebraic calculation.

An *equation* uses the equal sign (=) to state that two mathematical expressions have the same numerical value.

An *inequality* states that two expressions do not have the same value. The inequality symbols are called "less than" (<) and "greater than" (>), and the open end always goes toward the greater value. Think of it as a fisherman describing his catch—the wider his hands, the larger the fish.

box of the next line on the gameboard (or beside the next line of the scoresheet). Then do the indicated calculation:

- If your card is a spade or heart, add the value to each player's current score.
- If your card is a club or diamond, multiply each player's score by that number.

Write the new sum or product in each player's column on the scoresheet. Finally, draw the correct inequality symbol in the middle column.

A hand consists of two turns for each player, beginning with the non-dealer. So the dealer takes the next turn, playing a card, writing it at the start of the third line, and doing the calculations based on the values in the previous line. The non-dealer fills the next line, and the dealer plays the final line of that hand. Players do not draw new cards after each turn, so plan ahead how best to use the cards you have.

After the dealer's second turn, whoever has the greater value wins that hand. Circle the winning score. Mix all cards back into the deck and pass it to the other player to deal the next hand.

The first player to win three hands wins the game.

Sample Game

Tony challenges Steve to a game of Greater Than. Tony deals the first hand, and the players reveal their initial cards. Tony plays the queen of hearts, for a value of −10. Steve has the eight of spades, for a value of +8.

As the non-dealer, Steve goes first. He plays the three of clubs and writes "×3" on the next row. Then he multiplies both scores by three and writes each product in that player's column. He writes the less-than sign "<" to show that he's in the lead.

Tony lays down the ten of spades, adding 10 to each score. Steve chooses the five of hearts, adding −5 to the scores. Steve still has the

greater value, but Tony gets one more turn.

Tony plays the two of diamonds, which multiplies both scores by −2. Multiplying by a negative number changes the sign of the scores, and Tony wins the hand with 50 points.

Variations

HOUSE RULE: Does having the last play offer too great an advantage? Give the dealer one less card at the start of each hand.

Or add a level of risk to the game by having both players reveal their operation cards at the same time, as they did with their initial numbers. On each turn, choose a card and hold it face down until both players are ready. The card with the lesser value does its operation first. (And remember that −9 is less than +2.) This means the dealer may play before the non-dealer, depending on which cards they choose.

History

John Golden, who created this game, writes:

> *"I think this is a good educational game, but only close to a good strategy game. I can't quite figure out what's missing, so if you have a variation or adaptation to try, please let me know."*

If you think of a variation to share, post it in the comments on his original Math Hombre blog post.[†]

[†] *mathhombre.blogspot.com/2012/09/greater-than.html*

$$1 + 1 = ②$$
$$1 + 1 = ⑤ - 3$$
$$①+ 1 = \frac{200 - 3}{40}$$
$$\sqrt{⑩⓪} - 9 + 1 = \frac{200 - 3}{40}$$
$$\sqrt{10^{②}} - 9 + 1 = \frac{200 - 3}{40}$$
$$\sqrt{10^{(1+1)}} - 9 + 1 = \frac{200 - 3}{40}$$

A few rounds of the Substitution Game.
Players circled the number they wanted to change
and wrote their substitution below.
Then they copied the unchanged parts of the equation.

Substitution Game

MATH CONCEPTS: addition, subtraction, multiplication, division, order of operations, integers, fractions, equivalence and substitution.
PLAYERS: any number (a cooperative game).
EQUIPMENT: whiteboard and markers (preferred) or pencil and paper to share. Calculator optional.

Level One

The first player writes a simple equation at the top of the paper, such as "$1 + 1 = 2$." Then all players take turns complexifying this equation.

On your turn, copy the equation to the next line, replacing one number with an equivalent expression. For instance, replace the number 2 with:

$$5 - 3$$
$$\text{or } 50 \div 25$$
$$\text{or } (1/3) \times 6$$

… or any other calculation that equals two.

Use parentheses or brackets as needed to make your expression perfectly clear. For example, I put parentheses around my fraction above so people can tell I didn't mean "$1/(3 \times 6)$," which is definitely not a substitute for two.

If you have colored pencils or markers, circle the number you plan to substitute. Then write the substitution below in the same color. Finally, fill out the rest of the equation using plain pencil or a black marker.

The other players should check to make sure they agree with your math.

After you change part of the equation, it is no longer available for anyone else to use. If you substitute "$5 - 3$" for the 2, the other players cannot replace your creation with their own version of that number.

Words to Know

Two mathematical expressions are *equivalent* if they both have the same numerical value. You may *substitute* (replace) a number, variable, or expression with any of its equivalents whenever you like.

To *simplify* an expression is to substitute equivalent values that make it easier to read. Often this means trading in smaller numbers, as when we put a fraction into lowest terms (sometimes called "simplest form"). In algebra, simplification is often a matter of opinion, and the form you prefer depends on what you want to do with your expression.

If you get frustrated by mistakes and wrong answers in the Level Three simplification process, it's time to study the *order of operations*. (See the Operations game, next.) That's the most common source of confusion, especially when expressions include subtraction or division.

But they can alter individual numbers within your creation. So the next player may decide to substitute for the 5, writing a new expression in its place.

Continue until the paper is full, or until the equation looks satisfyingly complex, or until you run out of time. Save the paper (or copy the final equation from the whiteboard) for playing Level Three.

Level Two

As children grow comfortable with the basic game, pose additional challenges. Perhaps each substitution must use multiplication, or a fraction, or contain a specific number. If you're playing with a mixed-ability group, each player may have a different challenge.

Or if you've played Operations (see next game), you can use the card deck you made for that. Deal three cards to each player. On your turn, lay down one card from your hand. Your new expression must contain that operation. Draw a new card to replenish your hand.

For example, if you play the "÷5" card, you might replace the number 3 in an existing equation with:

$$15 \div 5$$
$$\text{or } (50 \div 5) - 7$$
$$\text{or } (\tfrac{1}{4}) \times (300 \div 5) \div 5$$

... or any other calculation that uses the operation on your card. Use parentheses or brackets as needed to make your expression perfectly clear.

The other players should check to make sure they agree with your math.

Continue until the paper is full, or until the equation looks satisfyingly complex, or until you run out of time. Save the paper (or copy the final equation from the whiteboard) for playing Level Three.

Level Three

The first player chooses an old round of the Substitution Game and writes the final equation at the top of the paper or whiteboard.

Players take turns simplifying the equation.

On your turn, copy the equation to the next line, replacing part of it with a simpler expression. The other players should check to make sure they agree with your math.

Continue until you reach the simplest form of the equation—which may not be the same as what the original game started with. For example, the simplest form of the equation "1 + 1 = 2" would be "2 = 2."

If the final equation is a true statement, then you win. Hooray!

But if anyone made a mistake in either level of play—either complexifying or simplifying the equation—you may end up with a nonsense statement like "2 = 13." Don't worry about trying to find the error. All the players still did a lot of mathematical thinking, so count this as a sideways win. Enjoy a good laugh at your silly result.

Variation

You can also play this as a solitaire game, just for the fun of complexifying equations. How crazy can you make the math?

History

Homeschooler Sonya Post's Arithmophobia No More website equips parents to make sense of math, so they can learn to teach their children. About the Substitution Game, she writes:

> "The student is learning that there's nearly an infinite number of ways to write any number. Students are learning the 'secrets' to building complex mathematical statements. You know what math a child has mastered when she can generate the math herself."

Operations

MATH CONCEPTS: addition, subtraction, multiplication, division, order of operations, integers, fractions.
PLAYERS: two or more.
EQUIPMENT: regular playing cards, blank index cards or card-sized pieces of paper, pencils or markers. Calculator optional.

Set-Up

Give each player a stack of ten index cards or blank squares of paper. On each card, write the symbol for an arithmetic operation (+, −, ×, ÷) and an integer. For example, you may create cards that say "+7" or "÷12" or "×(−2)." Make all of your cards different, but it doesn't matter if more than one player creates the same card.

Combine all the players' operation cards together in one deck. Save these cards for future games, adding a few more cards each time you play.

Agree on which color of playing card represents negative numbers. Aces count as one, face cards as twelve. (To make division easier, we skip eleven.)

How to Play

Keep the two decks of cards separate. Set the playing cards face down on the table. For the initial play, deal five operation cards to each player.

Then the dealer calls "high" or "low" for which value wins that hand. Finally, turn up the top card from the playing card deck to determine that hand's starting number N.

Players choose three operation cards from their hands and lay them down in sequence to form a mathematical expression beginning with the number N. Players each calculate the value of their own expression, following the standard order of operations.

For example, if the starting number N = 5, a player might arrange the cards +7, ÷12, and ×(−2) to create expressions like:

$$5 + 7 \div 12 \times (-2)$$
$$= 5 + {}^{7}/_{12} \times (-2)$$
$$= 5 + (-{}^{14}/_{12})$$
$$= 3\, {}^{5}/_{6}$$

or

$$5 \times (-2) + 7 \div 12$$
$$= -10 + {}^{7}/_{12}$$
$$= -9\, {}^{5}/_{12}$$

or

$$5 \div 12 + 7 \times (-2)$$
$$= {}^{5}/_{12} - 14$$
$$= -13\, {}^{7}/_{12}$$

etc.

The expression with the greatest (or least) value scores one point. Other players may want to examine the winner's calculation to make sure they agree, because order of operations can be confusing even for adults. If more than one player has the same winning value, they each get a point.

All players discard their used operations, keeping their two unplayed cards for the next hand. Pass the operations deck to a new dealer, who gives each player three cards to replenish their hands. If there aren't enough operation cards left, shuffle the discards back into the deck before dealing.

Turns continue until everyone has a chance to deal or until players agree to stop. Whoever has the highest score wins the game. In case of tied scores, players may deal one more hand as a final showdown.

Words to Know

The *order of operations* is a set of rules that govern the meaning of written arithmetic and algebraic expressions. Many of the "everybody gets this wrong" math memes on social media rely on confusion about order of operations, but these rules shouldn't be a "gotcha!" joke. They're a tool to help us communicate about math.

The basic principle is to do the most powerful operations before the weaker ones. Exponents are stronger than multiplication, for instance, and multiplication is stronger than subtraction. This hierarchy of strength creates the order of operations.

The first rule is that *grouping symbols* (parentheses and brackets) take precedence. We can use grouping symbols to show the part of a calculation we want done first, even if the normal order of operations would leave it until later.

One common area of confusion involves the fraction bar, which serves as an automatic grouping symbol. The numerator is one group, and the denominator is another. When entering complex fractions on a calculator, however, the fraction bar is not enough: You need to surround a complicated numerator or denominator with parentheses.

Whenever you're not sure whether people (or calculators) will understand the math you write, add plenty of grouping symbols to make your meaning clear.

Next in the order of operations are *the powers: exponents and roots*. Keep in mind that exponents only apply to the number they directly touch. So if you want to square a whole expression—or even a simple negative number with its minus sign—you need to group it with parentheses before writing the exponent.

$$-3^2 = -(3 \times 3) = -9$$
$$\text{but}$$
$$(-3)^2 = (-3) \times (-3) = 9$$

If you wish, you may allow exponent cards in your operations deck. Use the caret symbol "^" (which means "to the power of") and an integer.

Then comes *multiplication and its inverse operation, division*. These have equal strength under the order of operations. Unless grouping symbols get in the way, work them from left to right, like reading words in a book. But be careful not to let addition or subtraction fool you during this step. It's easy to get confused, especially when working math in your head.

The final stage in the order of operations features *addition and its inverse operation, subtraction*. Again, these operations have equal strength. Unless grouping symbols interfere, work them from left to right, like reading words in a book. Beware that your mind may try to trick you into adding first because it's easier. Work carefully, and everything will fall into place.

Variation

House Rule: Do your children want grouping symbols to add variety to their expressions? Create a set of parenthesis and bracket cards so a player can turn an expression like:

$$5 + 7 \div 12 \times (-2)$$
… into a friendlier calculation such as:
$$[5 + 7] \div 12 \times (-2)$$

Set these handy cards on the table where all players may reach them as needed.

History

Teacher Don Steward shared a series of "three operations" puzzles on his Median blog. I loved the puzzles and decided to turn his idea into a game.

Words to Know

In daily life, to *function* is to do something, and the function of an object or machine is the task it carries out. For example, a waffle maker's function is to transform batter into a deliciously toasted treat. But a *mathematical function* does not transform the input number—it simply matches each input with the defined partner output number.

Here is an example of a function rule:

$$\text{Double the number and then add seven.}$$
$$f(x) = 2x + 7$$

[Read that as "F of X equals…"]

If we give that function the input number $x = 5$, it finds the output partner $f(5) = 17$. Read this as "F of 5 equals 17."

We can write the input and output numbers as the ordered pair (5,17), joined inside a single set of parentheses. An *ordered pair* means two numbers connected in a specified order, usually (input, output) or for coordinate graphing (x,y).

The *domain* of a function is the set of allowable input numbers, and the *range* is the set of potential outputs.

When we write a function like $f(x)$, we use *variables*—letters or symbols that stand in for some other value. The variable f names the function itself. The variable inside the parentheses (x) represents the input number. Finally, the algebraic expression after the equal sign tells the rule we will use to find the matching output number.

Functions do not have to be named f, and inputs do not have to be called x. Those are the most commonly used generic variables, but you may choose any letters or symbols you like.

Function Machine

MATH CONCEPTS: addition, subtraction, multiplication, division, integers, fractions, absolute value, rounding, number properties, problem solving.
PLAYERS: two or more.
EQUIPMENT: pencils and paper for keeping track of input/output numbers. Calculator optional.

Set-Up

Every player needs pencil and paper. One person takes the role of the Function Machine, making up a rule for calculating with whatever numbers the other players provide. That person should write the function rule on a hidden slip of paper for reference as needed.

All the other players make themselves an input/output chart to keep track of the game.

How to Play

Each player in turn says a number that hasn't yet been chosen, and all players write that number in their Input columns. The Function Machine calculates with that number—mentally or on scratch paper *but not aloud*—and says the matching answer. All players write that number in their Output columns.

If the input number is too difficult, the Function Machine may say, "That's too hard," and ask for a different number.

Occasionally, players give an input number that makes no sense with the chosen rule. The Function Machine can say "That number is not in my function's domain."

An input/output chart, ready to play.

If you think you know the function rule, you must wait for your turn. When you give your input number, say, "I predict the output will be ___." If the Function Machine confirms your prediction, then you're allowed to guess the function rule.

The first player to identify correctly the function rule wins that round and becomes the new Function Machine.

Teaching Tip

My co-op math students have always enjoyed this game, especially when it is their turn to know the secret rule. But some kids freeze mentally at the task of creating a function rule, while others make up rules that are so complex they are nearly impossible to guess.

So I make a list of suggested function rules in advance, writing them on index cards or card-sized pieces of paper. Then I let the Function Machine players draw three cards and choose their favorite.

Here are some rules we have used:

- Add fifteen to the number: $f(x) = x + 15$.
- Subtract four from the number: $f(x) = x - 4$.
- Multiply the number by eight: $f(x) = 8x$.
- Cut the number in half: $f(x) = x \div 2$.
- Cut the number in half and then add one: $f(x) = (x \div 2) + 1$.
- Double the number and then subtract three: $f(x) = 2x - 3$.
- Add the next larger number: $f(x) = x + (x + 1)$.
- Multiply by the next larger number: $f(x) = x \times (x + 1)$.
- Triple the next larger number: $f(x) = 3(x + 1)$.

In math, a *function* is any rule that takes an input number and matches it with a specific output.

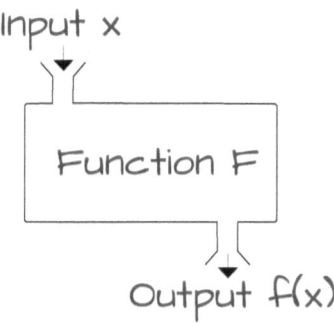

- Subtract the number from twenty-five: $f(x) = 25 - x$.
- Square the number and then add one: $f(x) = x^2 + 1$.
- Divide the number by three and say the remainder.
- Say the largest multiple of seven that is less than the number.
- Round off to the nearest hundred.
- Double the odd numbers, but cut the even numbers in half.
- All odd numbers match an output of seventeen, but all even numbers match twelve.
- Say the ones digit of the number.
- Add the digits in the number together.
- If the number is prime, say the number. If the number is not prime, say "one." (Remember that one, zero, fractions, and negative numbers are not prime.)
- Say the tenths digit of the number (the first digit after the decimal place). For whole numbers, the tenths digit is zero. The Function Machine player may need a calculator to convert fractional inputs to decimal numbers.

Be warned that some function rules can be described in more than one way. For example, the rule:

> Double the next higher number.

… could also be written as:

> Double the number and then add two.

The player who tries to guess the rule does not have to put it in the exact words the Function Machine used, as long as the statements are equivalent.

If you are playing with younger students, it helps to have a referee who knows algebra to judge the guesses. "Double the number, and then add two" is written as $2x + 2$, while "double the next higher number" is $2(x + 1)$. Anyone who knows algebra can see these are the same.

Words to Know

We have already defined expressions, variables, and substitution. Now it's time to put them all together.

To *evaluate* an algebraic expression means to choose values for the variables, substitute those values in the expression wherever each variable appears, and then do the resulting calculation. The answer to that calculation is the *value* of the expression when the variables are those numbers.

For example, evaluate $3x$ when $x = 7$:

$$3x = 3(7) = 21$$

To *solve* an algebraic equation or inequality means to find all the values for the variables that, when you substitute them in, make the equation or inequality true. Sometimes there is only one value that fits. Other times you can find a few values that work, and sometimes an infinite set of values.

And occasionally, you discover that *any* value for the variables makes the equation or inequality true. For example:

$$2y = y + y$$

Equations or inequalities that are true for all possible values of the variable are called *identities*. Can you think of another example?

Can you write an equation or inequality that will never be true, no matter what values we use for x or y? When equations like that happen by accident, we call them *mistakes*, but looking for them on purpose can be a fun.

Algebra Match

MATH CONCEPTS: variables, substitution, algebraic equations, inequalities.
PLAYERS: two to four.
EQUIPMENT: blank index cards or card-sized pieces of paper, two six-sided dice, pencils or markers.

Set-Up

Share out sixteen index cards or blank squares of paper among the players. Players write an algebraic equation or inequality on each card, as follows:

- ♦ Use only two variables, x and y. Each card must use at least one of these variables.

- ♦ The values of x and y will vary from 1 to 6, based on a dice roll. Make sure your equation or inequality makes sense in this range.

- ♦ Create equations and inequalities that work with more than one value for x and y. For example, a card with "$x = y$" matches any roll of doubles.

Label one additional card "X" and another "Y." Save all these cards for future play, adding a few new cards each game to create an Algebra Match deck.

How to Play

Lay all the algebra cards face up in a 4 × 4 array. If you have been collecting Algebra Match cards, shuffle your deck and turn up sixteen cards for the array. Pass the X and Y cards along with the dice for each player's turn.

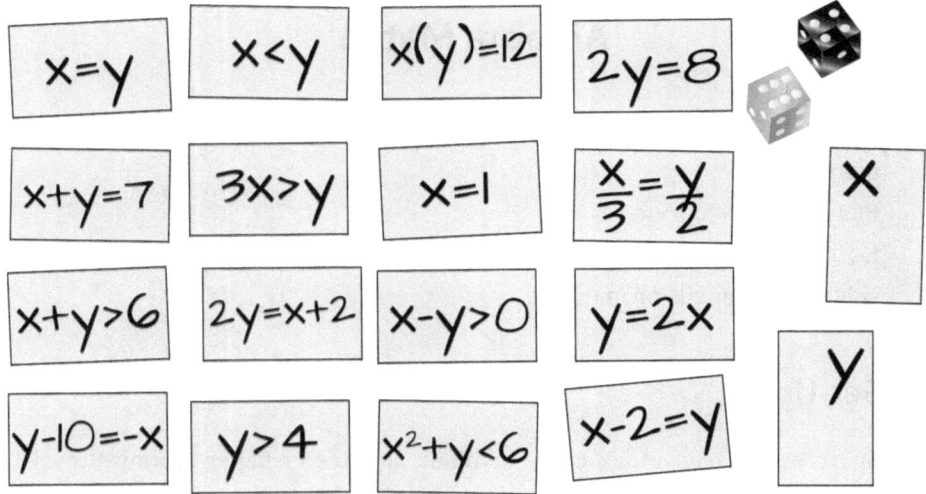

When you roll dice for *x* and *y*, which cards are the easiest to match? What would you take with this roll?

On your turn, roll the dice. Find an equation or inequality card that fits your numbers. Choose one die to be your *x* value and place it on the X card. Place the other die on the Y card.

Explain how you know the card matches your dice, and then add that card to your personal score pile. You can only take one card per turn, even if more than one matches your dice.

If you cannot find a matching equation or inequality card, say "Pass." But if another player sees a match you missed, they may claim it. If you placed the dice on the X and Y cards before passing, the other player may switch them to make a match. With more than two players, the first one to speak takes the missed card.

The game ends when all sixteen cards are claimed or when three turns in a row end with a pass. The player who collected the most cards wins.

Variations

ALGEBRA MULTI-MATCH: When you have collected a large deck of Algebra Match cards, allow players to claim up to three matching cards at a time. Place new cards from the deck to fill the spaces in your 4 × 4 array after each turn.

SOLITAIRE MULTI-MATCH: On each roll of the dice, claim every card that fits your x and y values. How many turns does it take to collect them all?

ALGEBRA-TAC-TOE: This requires just two players or two teams. You'll need a deck of at least thirty-six Algebra Match cards, plus two colors of poker chips or other tokens. Arrange the cards face up in a 6 × 6 array. On your turn, place a chip on one card that matches your dice roll. (Make a house rule: Do you want to allow both players to set chips on the same card?) The first player to claim four cards in a row wins.

History

This one is adapted from an algebra game by French homeschoolers Allison Carmichael and Martin Woods at the Parent Concept website. Visit their site for printable cards, if you don't want to make your own.[†]

[†] *parentconcept.com/printable-algebra-game*

Exhaust the Relationships

MATH CONCEPTS: addition, subtraction, multiplication, division, fractions, variables, equivalence and substitution.
PLAYERS: any number (a cooperative game).
EQUIPMENT: pencils and paper, or whiteboard and dry-erase markers. Cuisenaire rods, Legos, or other math blocks (optional).

Set-Up

Choose blocks of different colors and lengths to represent algebraic variables. Try to find blocks that have a relatively simple length relationship, such as "one block is twice as long as the other." If you have more than one block the same color, they should also have the same length.

Make two or three same-length rows of blocks. This models an algebraic equation because every row is equal to each of the other rows. If you don't have blocks to arrange, you can draw a similar diagram with colored rectangles.

For example, with Cuisenaire rods, we might build the following pattern:

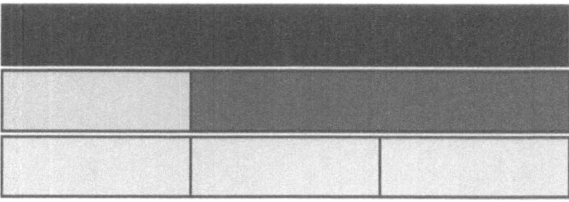

blue
light green + dark green
three of the light green

Give each size of block a variable name, such as G for light green and D for dark green. Variable names may be capital letters or lower case, your choice. On your paper, make a chart with one column for each size, labeled with the appropriate variable name.

How to Play

Take turns noticing and naming relationships between your blocks. Write an expression for each relationship on your chart.

In my picture above, there are only three basic relationships:

- ♦ Two light green rods are the same length as one dark green.
- ♦ Three light greens are the length of a blue rod.
- ♦ The difference between the lengths of blue and dark green is one light green rod.

But as you will see, there are nearly infinite ways to describe these relationships using mathematical symbols.

Begin by writing the simplest patterns. For example, my dark green rod is twice as long as the light green one. So in the D column, I might list two expressions:

$$G + G$$
$$2G$$

Look at the relationship from the other direction: The light green rod is half as long as the dark green one. And the difference between dark and light green is the length of one light green. So under the G label, I can list:

$$\frac{1}{2} \times D \text{ [or simply } \frac{1}{2}D\text{]}$$
$$D - G$$

Or looking at the blue rod, I may write these combinations in the B column:

$$G + D$$
$$3G$$

Can you see how I could describe these same relationships from the perspective of the green or dark green rods?

$$D = B - G$$
$$G = B - D$$
$$G = \tfrac{1}{3} \times B \ [\text{or} = \tfrac{1}{3} B]$$

G	D	B
$\tfrac{1}{2}D$	$G+G$	$G+D$
$D-G$	$2G$	$3G$
$B-D$	$B-G$	
$\tfrac{1}{3}B$		

These are the basic relationships between my three rods.

Complexify the Relationships

When you have exhausted the basic links between your blocks, it's time to play. Algebra is all about playing with ideas. So dig in and have fun.

You can keep adding new relationships to the columns of your chart. But as the expressions grow more complex, they take up a lot of space. I like to write my fanciest creations below the chart (or on a new sheet of paper) as equations.

For example, I might combine the following relationships:

$$D = G + G$$
$$D = 2G$$
$$G = \tfrac{1}{2} D$$

... to create any of these equations:

$$D = \tfrac{1}{2} D + \tfrac{1}{2} D$$
$$G = \tfrac{1}{2} (G + G)$$
$$D = \tfrac{1}{2} (G + G) + \tfrac{1}{2} (2G)$$

$$G = \tfrac{1}{2}[G + \tfrac{1}{2}(2G)]$$
$$G = \tfrac{1}{2} \times 2 \times \tfrac{1}{2} \times 2 \times \tfrac{1}{2} \times 2G$$
$$\text{etc.}$$

As players take turns writing expressions and equations, the others check for mistakes. If you're not sure, compare the algebra to the original blocks. Don't accept any math until you convince yourself it's true.

Build wilder and sillier algebraic monstrosities for as long as everyone wants to play, or until you run out of time. But keep in mind that short lessons work best in math. Thinking hard can be just as tiring as a physical workout.

You can always come back to your chart and add more ideas tomorrow.

Will you ever fully exhaust the relationships?

Variation

You can also play this as a solitaire game, just for the fun of complexifying equations. How crazy can you make the math?

History

Sonya Post, who shared this game on her Arithmophobia No More blog, writes:

> *"For those of you in the know, this looks very similar to the substitution game. Which, it is. Doing this activity, like playing the substitution game, helps the students see how all the operations and fractions are related to one another."*

Words to Know

Exponential calculations describe how things grow by multiplying, like bacteria in a culture or the interest on a loan—or a supervillain powering up to strike.

The *base* number is the type of growth: doubling, tripling, quadrupling, and so on. The *exponent* (also known as the *power* or the *index)* tells how many cycles of growth you have. And the value of an exponential expression tells how much *one* of the whatever-is-growing produces after that many growth cycles.

For example, if you roll a three while using base 4, you calculate three cycles of quadrupling growth.

$$4^3 = 1 \times 4 \times 4 \times 4 = 64$$

But if you roll one, it's only a single cycle of growth.

$$4^1 = 1 \times 4$$

The results of exponential calculations are called the *powers of the base number*. For example, the first five powers of two are 2, 4, 8, 16, and 32.

The base number can be less than one, in which case the "growth" is actually shrinking. Each multiplication cycle leaves you only a fraction of the amount you had before. Or imagine a negative base number, which creates a yo-yo-like growth that bounces up and down with each cycle.

(Exponents can also be negative, but that's harder to visualize. We'll deal with it later—see the Exponent Pickle game.)

Power Up

MATH CONCEPTS: powers (exponents).
PLAYERS: only two.
EQUIPMENT: one printed character sheet for each player, two six-sided dice, pencils or markers. Calculator optional.

Set-Up

Students create their own superheroes or supervillains using character sheets from the printables file or on regular notebook paper. Give your character a name and three superpowers, magical abilities, or technological gadgets. Describe the character's powers:

$$\text{Greater Ability} = \ldots$$

$$\text{Lesser Ability} = \ldots$$

$$\text{Random Ability} = \ldots$$

Briefly explain your character's origin story and draw a picture (optional).

How to Play

Each battle consists of three rounds, one for each of the superpowers. In the first round, players may choose which ability they want to use. For the second round, choose either unused power. Finally, the players call upon their remaining abilities.

Each player rolls one or two dice, as needed for their chosen superpower:

- Greater Ability: Roll one die. Use 4 as the base, your roll as the exponent.

- Lesser Ability: Roll one die. Use 3 as the base, your roll as the exponent.

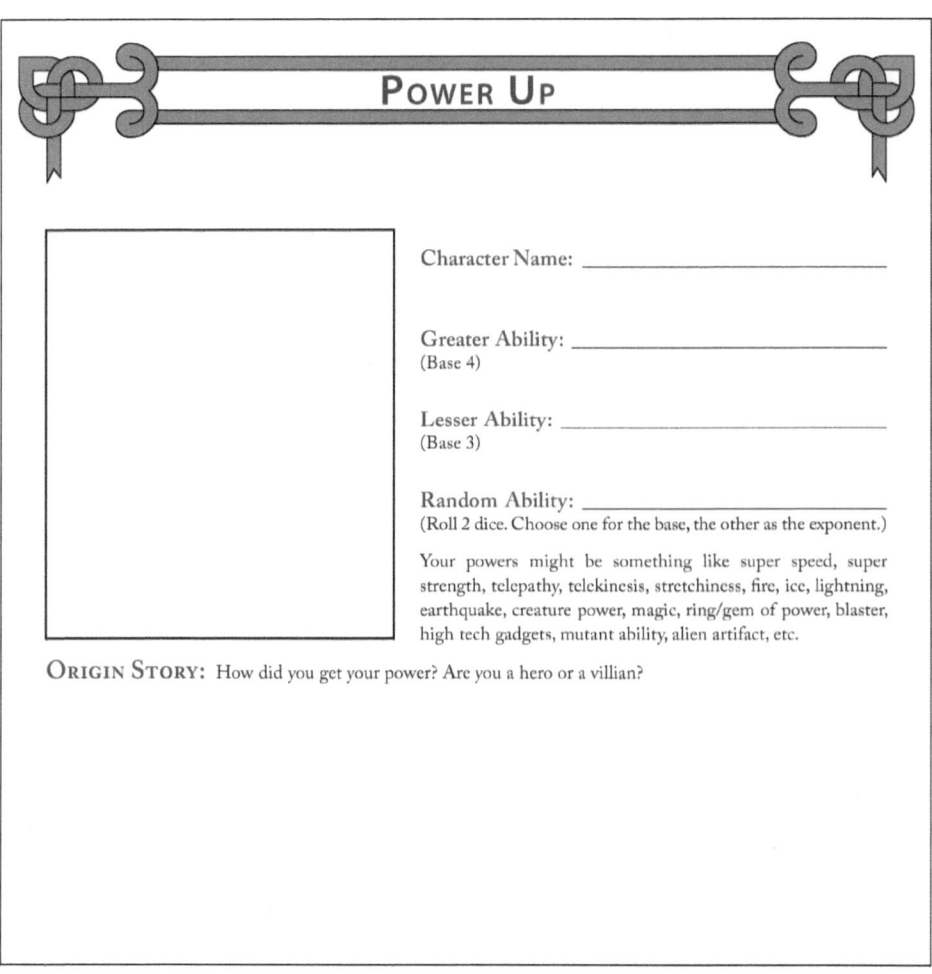

Print one Power Up character sheet for each player.

- Random Ability: Roll two dice. Choose which is the base and which is the exponent.

Calculate how much damage your attack inflicts upon your opponent. Whoever does the most damage wins that round.

The player who wins two of the three rounds wins the battle. If there is a tie, fight one more round, with both players using their random ability.

Variations

Players get twelve base points to split between their character's superpowers. For example, you might give one ability base 6 and both of the others base 3. A random ability (rolling two dice) costs four base points, so you could make all three superpowers random.

POWER TEAM ASSEMBLE: Two teams of players battle each other. For each round, teammates multiply their individual scores to calculate the total damage they inflict.

History

Game creator John Golden writes:

> "Note that it's okay for heroes to fight heroes, as those kinds of misunderstandings happen all the time. Villains might fight another villain because they're just so darn evil. (Or misunderstood.) Also, it's okay to end in a tie: 'We may have fought to a standstill, but I'll get you next time, Snake Lady!'"

Read the delightful origin story for this game—including a cartoon battle between Captain Victory and the Snake Lady—at Golden's Math Hombre blog.[†]

[†] *mathhombre.blogspot.com/2011/05/power-up.html*

Words to Know

Why is any number to the *zero power* equal to one? Think of exponential calculations as growth cycles. If you start with one of whatever-is-growing, and it goes through zero cycles of growth, how much does it change?

It doesn't matter what the growth pattern is, whether the item would be doubling or quintupling or anything. With zero growth cycles, the initial amount doesn't change at all. It is still one.

$$4^0 = 1 \times [\text{no growth}] = 1$$

It's as if the zero exponent marks the beginning of time, the starting position, right before the first cycle of exponential growth.

Exponent Number Train

MATH CONCEPTS: powers (exponents), estimation.
PLAYERS: two to four.
EQUIPMENT: one set of dominoes. Calculator optional.

How to Play

Turn all the domino tiles face down on the table and mix them around to make the woodpile. Each player draws six tiles from the woodpile but *does not look at them.* Players arrange their tiles in a row (train), as shown below. When all players are ready, turn the tiles in your train face up without changing which side is at the top.

Estimate the value of the exponential expression represented by each tile. The bottom number (closest to you) is the base, and the top is the exponent. Your goal is to make the values in your number train increase from left to right, but of course it will be mixed up to start with.

On your turn, draw one tile and decide which way you want to turn it to create an exponential expression with the numbers on the two halves of the domino. Then you have three choices:

♦ Use the new tile to replace one of the domino tiles in your train. Then discard the old one, mixing it into the woodpile.

♦ If you don't want to use the new tile, put it back and mix up the woodpile.

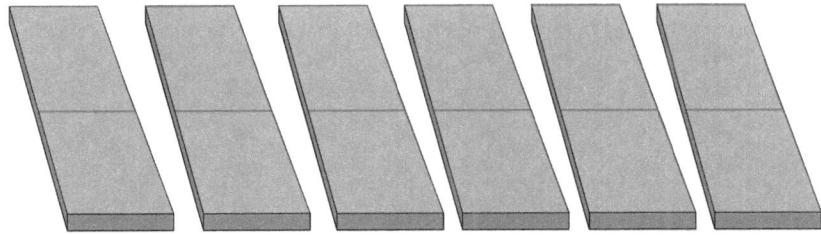

A domino number train, ready to flip and play.

♦ *Instead of drawing a new tile,* you may use your turn to rotate one of your current tiles, switching the base and exponent.

The first player to complete a train that increases in value from left to right wins the game.

Variations

Play with a different number of tiles. Try a shorter train for a quick game, or longer for a greater challenge.

HOUSE RULE: Decide how strict you will be about the "increases from left to right" rule and repeated numbers. Do you consider equivalent values (such as 1^6 and 3^0, or 4^2 and 2^4) as part of a valid train? Or must the player keep trying for a domino to replace one of the equivalents?

Exponent Pickle

MATH CONCEPTS: integers, powers (exponents), orders of magnitude.
PLAYERS: any number.
EQUIPMENT: deck of playing cards, whiteboard and markers for each player, or pencils and blank paper. Calculator recommended.

Set-Up

Each player draws a path with ten spaces big enough to write in. Exponent Pickle paths may be utilitarian or creative: a simple row of boxes, a curvy chain of circles, a series of stair steps, a caterpillar of ovals with legs, or a string of flowers with open centers. But every path needs to have ten spaces with a clear beginning and end.

If you draw the paths on paper, you can laminate these drawings or slip them into sheet protectors for repeated play. But if you make a new drawing each time, then the game can express the students' personalities as their artistic skills develop.

Agree on which color of cards represent negative numbers. Aces are worth one, and face cards are zeros. Spread the cards out face down to make a fishing pond.

How to Play

Players may take turns, or everyone may draw cards at the same time. If the whole group plays at once, all players must wait until the others discard before drawing for the next turn.

Draw two cards from the fishing pond. If they are the same color, choose a third card. If they're still all the same, keep drawing until you get a card of the opposite color.

Now, arrange your cards to make a math expression with an exponent. If you only have two cards, use one number as the base and the other as exponent. With more than two cards, get creative. Use the

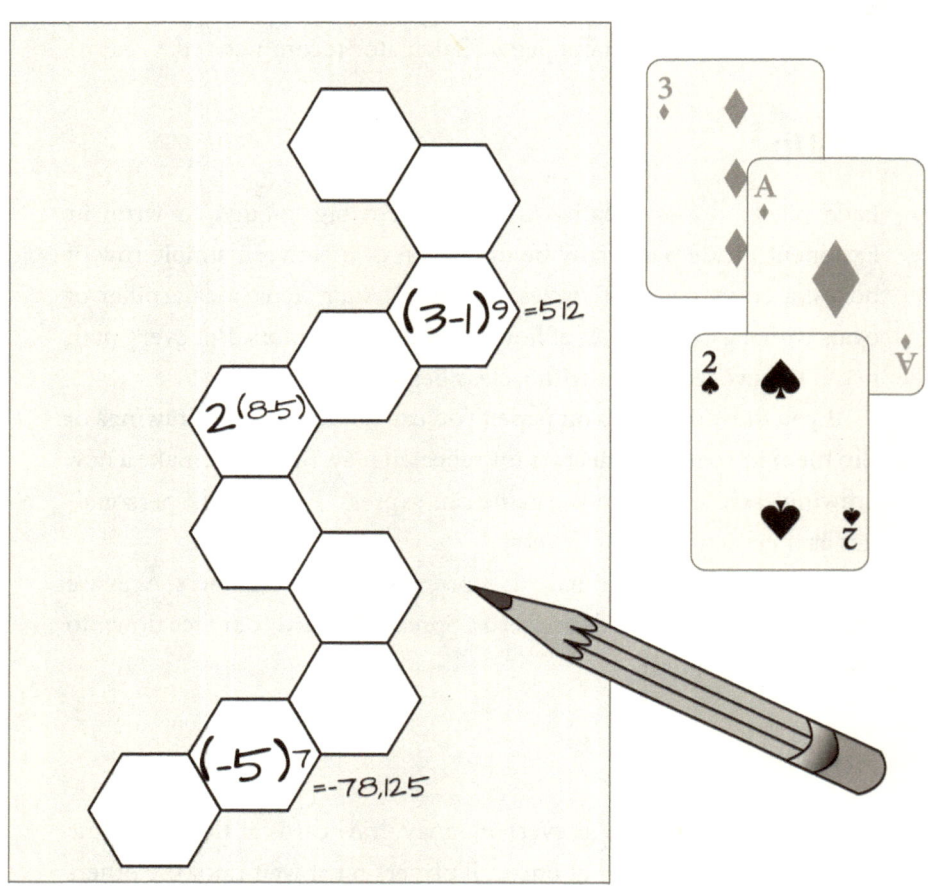

You can make $(-1 \times -3)^2$ or $2 \times (-3)^{(-1)}$ or $(-1/-3)^2$, and so on. Which number would you choose?

calculator to test your different possibilities.

Say your expression and write it into one space on your path. (Also write the value of your expression beside it, if you needed the calculator.) Always make sure the values increase from the beginning of your path to the end. If you can't create an expression that keeps the least-to-greatest pattern, then you miss that turn.

To end your turn, place all your cards face down into the fishing pond. Mix them in thoroughly, so the next player can't guess which card is where.

The first person to fill a path, with all the values in order from least to greatest, wins the game.

Variations

Exponent Pickle is also fun as a solitaire game. Can you fill your path without missing a turn?

FOR BEGINNERS: Play a simpler game. When players draw three cards, they must write an expression of this form:

$$A \times B^C$$

Lucky players who draw more cards may choose any three of the numbers to make an $A \times B^C$ expression.

HOUSE RULE 1: Decide how strict you will be about the "increasing order" rule and repeated values. Can a player use both $(-6)^0$ and $1^{(-5)}$ as part of a valid path? Or must the player keep trying for new cards to replace one of the equivalents?

HOUSE RULE 2: Do your children get bored waiting for everyone else to play? Let all players draw cards at once, so everyone has something to work with. Be sure to check each other's calculations before the next turn.

HOUSE RULE 3: Does a poor choice early in the game frustrate your students? Allow players to erase an expression from their paths, in place of taking a regular turn.

History

John Golden invented the simpler "$A \times B^C$" version of this game and tested it in his college algebra class. He writes:

"Older students also need these play experiences. I think they just abstract from them more quickly than younger students."

Words to Know

Why is taking a number to a *negative power* the same as dividing? Think of exponential calculations as growth cycles. For example, let's think of the powers of three. These grow by tripling every cycle.

In the starting position (the zero cycle). you have one of whatever-is-growing. But what if we imagine going back in time, one growth cycle *before* the zero position? That would be the −1st cycle. Its value must be some amount that when it grows one more cycle, it will equal one. What number is that?

$$3^{(-1)} \times 3 = 3^0 = 1$$
$$3^{(-1)} = 1/3$$

Now imagine going back to the −2nd cycle. The amount has to grow by two cycles to get up to the zero position. What value is that?

$$3^{(-2)} \times 3 \times 3 = 3^0 = 1$$
$$3^{(-2)} = 1/9$$

And so on. Each negative power is like a peek back in time. A negative power imagines growth cycles *before* the beginning of exponential growth. It's no wonder many early mathematicians considered negative numbers absurd.

Krypto

MATH CONCEPTS: addition, subtraction, multiplication, division, order of operations, fractions, decimals, integers, absolute value, powers and roots, factorials.

PLAYERS: any number.

EQUIPMENT: one deck of cards, pencil and paper for each player and for keeping score. Calculator optional.

Set-Up

Agree on which color represents negative numbers. Agree on the value of face cards (for example, jack is eleven, queen is twelve, and king is thirteen). Or maybe you want all face cards to equal ten, while the ace can be worth the player's choice of either one or eleven.

Decide whether to set limits on the math operations allowed in your game. Or let dealers set their own limits at the beginning of each hand, before turning up any cards.

Also, agree on how long you want to play. A traditional game of Krypto runs for ten hands, but you may prefer a shorter game.

How to Play

The dealer turns up one or two cards (dealer's choice) and arranges them to make the target number. For example, if the cards are 2 and 7, the target could be 27 or 72.

Leaving a space to separate them from the target number, the dealer turns up five additional cards. Players race to use these numbers in any order to create a mathematical expression equal to the target number.

You may add, subtract, multiply, or divide the numbers, and you can use a card as an exponent, but you cannot put two cards together to make a two-digit number. You must use all five cards. Each card may be used only once in the calculation.

Calculations may include:

- The arithmetic operations +, −, ×, ÷
- Decimal points, so that 5 could become 0.5 (The zero before a decimal point doesn't count as making a two-digit number. But you cannot add zeros after the decimal point to make something like 0.05.)
- Absolute values
- Square roots
- Exponents—take one card to the power of another
- Factorials
- Parentheses, brackets, or other grouping symbols
- The overhead-bar *(vinculum)* to mark a repeating decimal

You may combine these calculations in creative ways. For example, if one of the cards is −1, you could use a decimal point and a vinculum to make the repeating decimal equivalent to −1/9. Then use the absolute value, if a positive fraction is more useful in reaching the target number. And the square root can turn that positive fraction into 1/3, if you wish.

A player who figures out how to make the target number shouts "Krypto!" The player must demonstrate the calculation, either in writing or by moving the cards around, so the other players can check it.

If the calculation is valid, that player scores a point. If the same player calls Krypto on the next hand, he adds two points to his score. The point value doubles in each succeeding hand until another player breaks the winning streak. If the calculation was not valid, however, the player loses a point and must sit out the next hand. A player's score can go below zero.

If all players agree the hand is impossible, the dealer turns up three additional cards. Players may choose any five of the eight cards (the original five plus these three) to make the target number.

After each hand, the deal passes to the next player. The highest score after ten hands wins the game.

Variations

POWER KRYPTO: Do your children hate games that require speed? Let players use any of the cards to make their expression, writing it on paper and keeping it hidden until everyone else is ready. You may use each card only once, as in the original game, but you do not have to use all five cards. Then players all reveal their expressions. If you had time to think of more than one expression, reveal your favorite.

If the other players agree your expression works, count the number of cards used and score two to that power. And if your expression is unique, you score double points. For instance, suppose you used four of the five cards to calculate the target number, and no one else shared the same expression. You score $2^4 = 16$ doubled, for a total of 32 points.

The first player to reach 300 points wins the game. If more than one player passes 300 on the same turn, the highest score wins.

ROAD TRIP KRYPTO: Each player needs a clipboard (or other hard surface to write on) with paper and pencil. Players take turns naming a positive or negative number, until there are eight numbers named. Numbers may be repeated. All players write these game numbers on their paper.

Then the driver names a target number. The driver also sets a time limit for the game—perhaps until the next gas station or rest area. Players try to make the target using one or more of the game numbers as many ways as they can. Each game number may be used only once per calculation. But if the game numbers include duplicates, you can use both in the same expression.

Every valid expression equal to the target scores one point per number used. Unique expressions score double.

NUMBER YOGA: Turn up five cards and challenge yourself to find expressions for as many numbers as possible. How many values can you make?

THE YEAR GAME: Instead of playing with cards, use the digits of the current year. How many numbers can you make?

History

Mathematicians have played with calculation puzzles (like the Four Fours) for centuries. Authors Bill Lombard and Brad Fulton coined the phrase Number Yoga. Daniel Yovich invented Krypto in the 1960s and sold the game to Parker Brothers. The original version used a special deck of fifty-six cards and featured more restrictive calculation rules.

Krypto Insanity

MATH CONCEPTS: addition, subtraction, multiplication, division, order of operations, fractions, decimals, integers, absolute value, powers and roots, factorials.
PLAYERS: any number.
EQUIPMENT: one deck of cards, pencil and paper for each player and for keeping score. Calculator optional.

Set-Up

Agree on which color represents negative numbers. Agree on the value of face cards (for example, jack is eleven, queen is twelve, and king is thirteen). Or maybe you want all face cards to equal ten, while the ace can be worth the player's choice of either one or eleven.

Decide whether to set limits on the math operations allowed in your game.

How to Play

Deal ten *hands*—sets of five cards each, face down on the table. Turn up one of the two remaining cards as the first target number. Set the final card under it, to use later.

Play three rounds:

- ♦ Round 1: Current face-up target number
- ♦ Round 2: Final card as target number
- ♦ Round 3: Target number is zero

Everyone plays at once. Pick up any of the face-down hands. Try to make the target number with those five cards (using the same operations as in ordinary Krypto). If you cannot see a solution, turn that hand face down and pick up another. You may choose a different hand as many times as you like.

When you (or any player) call "Krypto," everyone else stops working to listen. Show your cards and explain the calculation. If all players agree it's a valid expression, set that hand of cards in your scoring pile. Then grab another hand and keep going. If your expression isn't valid, return that hand to the table and try a different set of cards.

Keep the hands in your scoring pile separate by turning each set of cards at a right angle to the set below it. That makes it easy to count points and keeps the hands ready for the next round.

If there are no hands left on the table, players can share with anyone who still has cards. On a shared hand, the first player to call Krypto scores the point.

After all the other cards are claimed, the player who has the last hand spreads it face up on the table. Everyone races for the final Krypto.

If all players agree it's impossible to make the target number with any hand, put those cards aside while you count score. Players score one point per hand in their scoring piles. Then return all ten sets of cards to the table, face down, to play the next round.

The player with the highest final score wins the game.

The Number That Must Not Be Named

MATH CONCEPTS: addition, subtraction, multiplication, division, integers, fractions, factoring, powers and roots, prime numbers, and other number properties.

PLAYERS: two or more (a cooperative game).

EQUIPMENT: none.

Set-Up

Because all calculations are done mentally, players must agree on what types of numbers are allowed. For example, beginners may want to start with the positive whole numbers 1–100. As players gain experience, you can expand the range of possibilities.

How to Play

The first player names any number within the permissible range. Players take turns naming mathematical operations, performing each calculation mentally but never saying their answer aloud.

For example, suppose the first player names "15." Turns may then proceed as follows, with the number changing as shown in parentheses:

"Times two." (30)

"Divided by five." (6)

"Squared." (36)

"Subtract it from one hundred." (64)

"Square root." (8)

"Cube root." (2)

"To the fifth power." (32)

"Plus one." (33)

"Nearest prime number." (31)

etc.

Players try to show style by naming operations that haven't been used, especially something particular to the current number. Since the last calculation left the number at thirty-one, you might say "plus sixty-nine." This proves you've been paying attention and gives everyone's brain a brief rest on the nice, round number 100.

If a player names a calculation that makes no sense or that takes the number outside the agreed-upon range, that player is out of the game.

At any time, one player may challenge another to name the current number. If the challenged player says the wrong number, that player drops out of the game. But if the answer is correct, then the challenger is out.

The game continues until only one player remains, or until the players decide to stop.

History

When I was a kid, our teachers used to make students keep up with a long chain of mental calculations. This game offers students a chance to fight back and see if they can stump the teacher. I found the game on Joel David Hamkins's blog. Your children may also enjoy his Rule-Making Game.[†]

[†] jdh.hamkins.org/the-rule-making-game

Geometry

I need to tell you a secret. I used to hate teaching geometry.

Geometry — in my former way of viewing it — had too much vocabulary, too much formality, too many rules.

But I have changed.

I now see geometry as a playground of ideas for students of all ages. Geometry now presents opportunities for exploration, wonder, and delight.

— Christopher Danielson

Activity: Star Polygons

MATH AND ART COMBINE BEAUTIFULLY to stimulate children's thinking about geometry. In this project, students will create a variety of star polygons and related figures to explore symmetry, angles, and number relationships.

Your children are probably used to thinking about convex polygons like triangles and squares. In a *convex* polygon, none of the corners "bump in." The official mathematical definition for a convex shape is that you can connect any two points with a line segment that stays entirely within the shape. Using that definition, can you prove that two of the figures below are *not* convex?

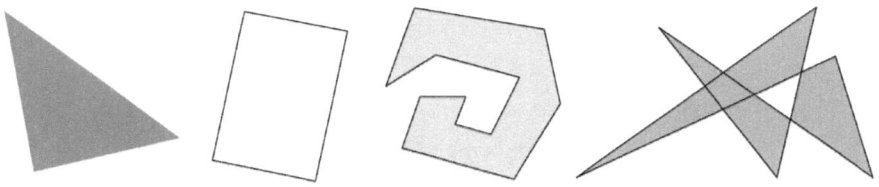

Polygons are flat, *closed* (connected all the way around) shapes with all straight sides.

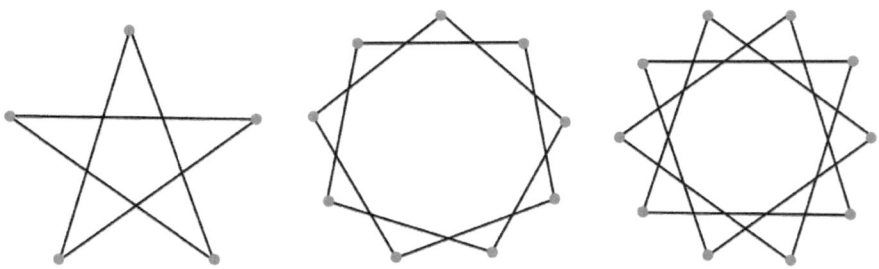

Star polygon edges cross without forming an intersection, as if they existed on different layers. The *vertices* (points of the star) are the only places where the line segments meet each other.

Students may also have worked with *concave* polygons, which have a "bumped-in" part. Or as mathematicians would say, concave polygons have at least one *reflex* (between 180° and 360°) interior angle.

Star polygons are complex, non-convex geometric figures in which line segments crisscross each other several times before connecting back to the starting point. The *pentagram,* that five-pointed star you can draw without lifting your pencil, is a star polygon.

Drawing Stars

Print several copies of the six-circle Star Polygons Template from the *Prealgebra & Geometry Printables* file for each student. Or let children draw their own circles.

Here is how to make a math star:

- ♦ Choose any (relatively small) whole number and call it n. Draw n dots as evenly spaced as you can around the circumference of your circle.

- ♦ Now choose any whole number less than n. That is your dot-counting number. Call this number k.

- ♦ Place your pencil on any dot. Count around the circle k dots. Draw a straight line segment from your starting position to that new dot.

- ♦ Count k more dots. Draw a new line from your current position to that next dot. Always count in the same direction on your circle.

- ♦ Continue to count k dots and draw connecting lines until you come back to where you started.

- ♦ Are there any dots left out, unconnected to your star? If so, lift your pencil and move to any unused dot. Start a new shape on the same circle, counting dots and drawing lines as before. Repeat as often as necessary.

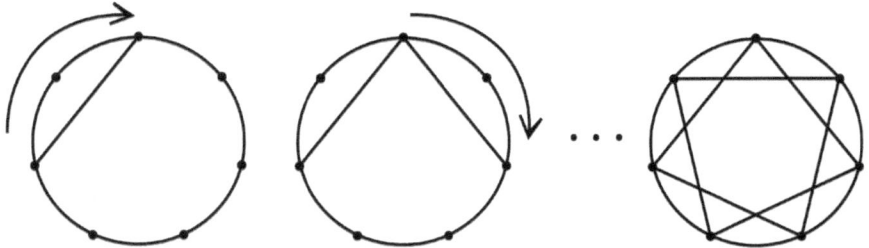

Stages in drawing the star polygon for $n = 7$ and $k = 2$.

- Now are all your dots included in a shape? Are your shapes closed? A *closed* shape connects all the way around, back to its starting point.
- Label your drawing: $\{n/k\}$, which you read as *"n count k."* The example above is the $\{7/2\}$ or "7 count 2" star polygon.

Some kids will take off and run with the project, drawing many stars to explore what happens. Others may lose interest quickly—if this describes your children, try asking for a few stars per day stretched over a week or two, gradually building a collection to study.

For students who enjoy gamification, let them roll dice to find random numbers for their stars. Use the greater number for n and the lesser for k.

What Do You Notice?

Take time to notice and wonder about the shapes students have drawn. This is the mathematical equivalent of a nature walk. There is no specific destination. Your goal for children is that they fully experience the journey.

Begin with a large assortment of star figures based on various numbers for n and k. Cut out your $\{n/k\}$ drawings to make star cards. Have your students sort the cards by whatever categories they like.

How many types of star figures can they discover?

Ask your children to list the things they notice about their designs. Think about things like these:

- Reflection symmetry: Does the shape have a mirror line? Does it have more than one?
- Rotational symmetry: Can the shape turn like a spinner? How many times does it match itself as it turns? (The starting/ending position counts as one match.)
- How many points are on the star? Are they fat obtuse angles? Or are they thin acute angles? Can a star have right angles?
- How many times did you go around the circle to make the star figure?
- What shapes do you see in your star?

You may want to begin with conversation. Talk with your students, taking turns to notice things out loud, before sending them off to write their lists.

For younger children or for those who struggle with writing, take dictation. You want to give them freedom to focus on their thinking, not on the mechanics of putting thoughts onto paper.

What Do You Wonder?

Start a new list for things your children wonder about. For example:

- If you space the dots irregularly around your circle, how does that change your stars?
- What would happen if you picked k equal to or greater than n?
- What if you chose a negative number for k?
- Could you use a fraction or mixed number?
- Could there be any combination of n and k numbers that make an infinite cycle, where your pencil never comes back to its starting dot?

- How would it look if you picked an *n* and drew all the *k* patterns on a single circle?

Experiment with the patterns you noticed. Draw new, different star figures to find out whether the pattern continues. Make conjectures and test them.

Create Math Art

Ask your students to choose a favorite star polygon and turn it into mathematical art. For instance, they might…

- Draw the polygon on a big circle, and color it. Cut it out to mount on a contrasting background.
- Doodle a pattern in each section of the star. Add bits of accent color and balancing areas of dark shading to create a pleasing design.
- Draw the star with white crayon, oil pastel, or masking fluid. Paint with watercolor.
- Use colored string or embroidery floss to stitch a star pattern on canvas or card stock.

In the printables file, I've included a large circle worksheet to draw stars for coloring, or you can encourage children to work even larger on art paper. Or work small—circle art can be charming on handmade greeting cards.

Explore Polygon Patterns

Depending on your children's ages and interest levels, you may want to stop after creating math art. Or continue to investigate the mathematical patterns your students noticed and the questions they wondered about.

The pattern chart on the next page may prompt new discussion.

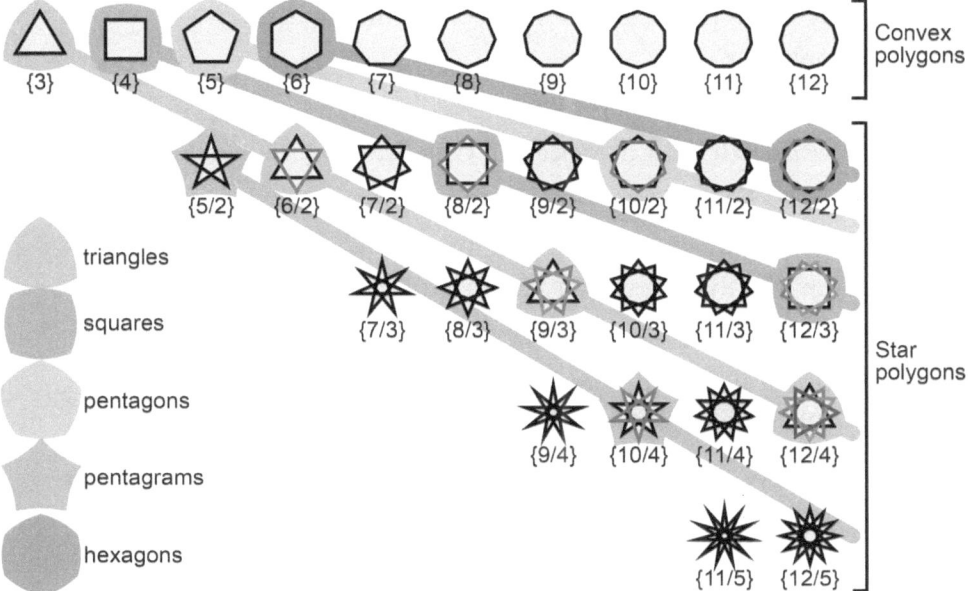

[Picture from Wikimedia Commons] This chart highlights several patterns in the shapes of mathematical stars.

- What relationships do you see?
- Does the chart make you wonder?
- Which of these star figures did you draw?
- What combination of *n* and *k* will you try next?

Study Your Stars

There is no precise geometric definition of a star shape. In this section, I use *star figure* as the general term, while *star polygon* refers to {*n*/*k*} figures that connect all their dots in a single, crisscrossing line.

If your students are ready for a deeper dive into star math, investigate these questions:

- Are there combinations of {*n*/*k*} that can't make any star figure? Why not?

- Which combinations of *n* and *k* form the regular convex polygons?
- Some star figures look like asterisks, with straight lines that cross each other in the center. What kind of {*n/k*} numbers create these shapes?
- Sometimes two designs using different {*n/k*} numbers end up with the same shape. How does that happen, and why?
- With some combinations of *n* and *k*, you must lift your pencil and make more than one shape to hit all the dots. What is special about the {*n/k*} numbers for these figures?
- Is it possible to predict whether {*n/k*} will form a true star polygon?
- Measure the angles at the points of your stars. Do you notice any pattern?
- Could you determine the angles in a star without measuring?

What other types of stars can you find in math? In art? Perhaps you'd like to explore some related patterns from Eric Broug's School of Islamic Geometric Design.[†]

[†] *sigd.org/resources/classroom-resources-islamic-geometric-design*

Penterimeter

MATH CONCEPTS: pentominoes, perimeter.
PLAYERS: two players or two teams.
EQUIPMENT: pentominoes or square grid paper, pencils, paper for keeping score.

Set-Up

You need one set of pentominoes for players to share. A *pentomino* (rhymes with *domino*) is a flat shape made of five squares where each square shares at least one complete edge with another square in the shape.

To make your own pentomino pieces, print a few sheets of square dotty or lined grid paper with medium-to-large squares on stiff paper or card stock.

Ask your children to color in five-square shapes that fit the definition above. Squares that meet only at a corner do not count. How many different pentominoes can they find?

The total number of pentominoes—twelve or eighteen—depends on whether you count the mirror-image pieces as different or as a

If you have store-bought pentominoes or pieces from a game like Blokus, you could use those. But students learn more by creating their own game pieces on square grid paper.

Words to Know

Many children find our next two vocabulary words confusing.

The *perimeter* is the sum of all the side edges of a flat shape—the distance all the way around. Your students can think of the word *rim* to remember perimeter, or think of building a fence to surround a field. Perimeter is a length measurement, using units like meters or feet.

The *area* is the amount of flat space covered by the shape, inside the perimeter. It is like the field inside the fence. Your children might think of "squarea" because area is measured by how many square units (like m² or ft²) it takes to fill the shape.

single piece flipped over. But if your children identify at least eight pentomino shapes, that's enough to start playing the game.

Cut out your paper pentominoes and place them on the table where all players can reach. Or buy small wooden cubes and stick them together with craft glue to make 3-D game pieces.

On another sheet of paper, make a score sheet with three columns: one for each player's score and one to keep track of your shape's growing perimeter.

How to Play

The first player chooses a starting pentomino and moves it to the middle of the table. Rather than counting the whole perimeter, they score one point. For the "Perimeter" column, write the distance around the shape, counting each side of a square as one unit.

On succeeding turns, you may choose any unused pentomino. Add it to the growing shape so that it matches alongside at least one edge of a square but doesn't cover any of the previous pieces.

Count your shape's new perimeter and determine how much it increased from the previous turn. Add that many points to your score. (If the perimeter decreased, that counts as a negative increase.)

Alternate turns until all the pentominoes are played or until there aren't enough pieces left for both players to have another turn. The player with the lowest total score wins.

Variations

If players want a greater challenge, print a new sheet of square grid paper to make a gameboard. Color in a few squares as obstacles. Pentominoes must line up with the gameboard grid, but they may not cover the colored squares.

GOLOMB'S GAME: Players take turns placing pentominoes within an 8 × 8 square gameboard. Pieces may not overlap, and they do not have to touch those already played. The last player to place a shape, leaving the opponent without a move, wins the game.

History

Penterimeter is another game from the prolific John Golden. He writes:

> *"It was pretty fun, and surprisingly strategic. Students were surprised you could score 0, and astonished when someone scored negative points. In the long run, I think the game gets repetitive, but it has given students a lot of experience with perimeter by then."*

Golomb's Game is based on a checkerboard conundrum from Henry Ernest Dudeney's 1907 book *The Canterbury Puzzles* and named for the modern mathematician who first studied polyominoes, Solomon W. Golomb.

Area Block

MATH CONCEPTS: area, square units, regular and irregular polygons.
PLAYERS: two players or two teams.
EQUIPMENT: square grid paper, two different colored pencils or markers, scratch paper for keeping score.

Set-Up

Outline a 10 × 10 square grid for your gameboard. After you get used to the game on a plain grid, get creative. Draw a larger gameboard and shade in enough "holes" to leave a playing area of approximately one hundred open squares.

For example, you could outline a 12 × 12 area and then color forty-four squares as off-limits. Or let players take turns shading out-of-bounds sections with a neutral color before play begins.

How to Play

Players take turns claiming area by coloring in squares. On the first turn, the first player may claim up to eight squares. In all succeeding turns for either player, the limit is ten squares.

Indulge your creativity designing gameboards for Area Block. Be sure to leave roughly one hundred squares open for players to claim.

Words to Know

A *regular* polygon looks the same all around. All the sides are the same length, and all the angles are the same size. For example, a square is a regular polygon, and so is the hexagonal cell of a honeycomb.

An *irregular* polygon has at least one side or angle different from the others. Most of the shapes you draw in Area Block will be irregular polygons.

On your turn, color squares that touch each other along their sides to make a solid polygon. You may color partial squares or make the perimeter of your shape slant at a diagonal, but your squares may not touch only at a corner. Your polygon does not have to touch an area you've already claimed.

Add the area of the polygon to your running total score.

When the gameboard is full, whoever has claimed the greater area wins. The loser gets to choose whether to play first or second in the next game.

Variations

AREA DICE: Add a touch of randomness to the game. On your turn, roll two six-sided dice. You may claim an area up to the sum of your dice numbers.

PERIMETER BLOCK: Color a polygon with an area up to ten squares, as in Area Block. But your score is the perimeter of your polygon. Polygons must follow the grid—no slanted edges allowed.

History

John Golden, who created this game, writes:

> "Blokus meets Nim, this game works on area and strategy. It's pretty fun!"

Area Battle

MATH CONCEPTS: area, square units, perimeter, regular and irregular polygons.
PLAYERS: two to four.
EQUIPMENT: homemade game cards, pencils or markers.

Set-Up

Students cut square grid paper into 5 × 5 squares. Each student creates a personal deck of eleven cards, as follows:

- Regular or irregular polygons, one with each shaded-in area: 1, 2, 3, …, 8, and 9 square units. The area must be easy to count, so build each polygon from a combination of squares and half-squares.
- Two additional polygon cards with areas of the player's choice.

Claim your deck: Mark the back of each card with your insignia (initials, emoticon, math symbol, etc.—your choice). Decorate as desired, so long as it doesn't bleed through the paper and obscure your polygons.

How to Play

Stack your card deck face down in front of you. Draw the top two cards for your hand.

On your turn, call "high" or "low" for whether the greatest or least area wins the trick. Then each player chooses a card from their hand and holds it face down. When all players are ready, turn the cards up, and the winner captures them all. Each player keeps a pile of prisoner cards.

In case of a tie, those players reveal their other card, and the longest perimeter takes the trick. If it's still a tie, leave the cards on the table for the winner of the next trick.

Players draw one or two new cards to replenish their hands before the next turn.

The game ends when at least one player runs out of cards. Whoever has captured the most prisoners wins the game. Give the prisoner cards back to their owner, unless you are playing the Spoils of War variation.

Variations

In between games, players can make new cards. But they must always play with a deck that fits the rules: nine cards with the areas 1–9, plus two extras.

PERIMETER BATTLE: On each turn, the low or high perimeter takes the trick. In case of a tie, the greatest area wins the second fight.

SPOILS OF WAR: If you win the game, you may claim one card from your prisoner pile to replace the matching card from your deck. Erase the identification and add your own symbol on the back. You may offer your old card to the losing player as a consolation prize, or keep it as a backup card.

A sample deck of Area Battle cards. Slanted edges may be tricky to count, if players need to compare perimeter in case of a tie. But that's okay—math debates prompt deeper thinking.

History

When John Golden created and tested this game, the students gave it "universal thumbs up as to whether they would recommend it to other teachers. They were requesting time to play in class later, which is surely a good sign."

Prism Power

MATH CONCEPTS: rectangular volume, cubic units.
PLAYERS: two or more.
EQUIPMENT: printed gameboard or plain paper, pencils or markers, one six-sided die, 40–50 cubic blocks per player.

Set-Up

Interlocking cubes work well for this game, as do plain wooden cubes. We've also played with sugar cubes when we found them cheap—but do spread a napkin or something to catch the inevitable sugar crumbs.

Print a gameboard for each player. Or make your own scoresheet with four columns labeled *length, width, height,* and *volume* (total number of cubes) and use a second sheet of paper for your building property. Draw a straight line to represent the street-facing edge of your property.

If you have enough dice, players may all take their turns at the same time.

How to Play

Each player starts with two blocks, creating a *rectangular prism* (box shape). Place the long edge of your initial prism parallel to the street-edge of your paper property. This side is your building's "length" throughout the game.

All players record their building's measurements for length (2), width (1), and height (1) and the initial volume (2 cubes). One side-edge of a block is one length unit, and the volume is the total number of blocks.

On your turn, roll the die and follow the instructions that match your number:

♦ 1 = Architect's Choice: Add one layer to any dimension.

♦ 2 = Elbow Room: Make your prism one layer wider.

Words to Know

What most people call a "box shape," mathematicians call a *right rectangular prism*—a prism with a rectangle for its base.

A *prism* is any shape with a polygonal *base*—actually two identical bases, the top and bottom—and flat *sides* that connect the two bases. All the side edges are parallel to each other. Think of a deck of cards with each card cut into an identical polygon shape. When you stack all the cards in the deck, you create a prism with that polygon as the base.

When the deck of cards is stacked straight up, we call that a *right* prism because the side edges make a right angle with the table. Each side of a right prism is a rectangle, no matter what shape is the base. The common glass or plastic prism used to turn light rays into rainbows is a right triangular prism.

If the deck of cards slants sideways, making each side a parallelogram, the prism is *oblique*. Notice that one deck of cards can form either a right or oblique prism—or several oblique prisms at different angles—but the *volume* (total space enclosed) never changes. Any oblique prism has the same volume as a right prism with the same base and height.

Incidentally, the idea of imagining a shape as a stack of cards is called *Cavalieri's Principle* (named after Bonaventura Cavalieri, one of Galileo's students). Your children will use this principle in calculus to find the volume of two- and three-dimensional objects.

- 3 = Expanding Storefront: Make your prism one layer longer.
- 4 = Penthouse Apartment: Make your prism one layer taller.
- 5 = Community Investment Grant: Increase the smallest dimension by one.
- 6 = Zoning Violation: Remove one layer from the dimension of your choice.

After each turn, record your building's new dimensions and volume on your scoresheet.

Notice that the street-facing length of your building may not always be the longest side, depending on how the dice roll. It often happens in algebra that the side we called "length" when we began a problem turns out to be shorter than the side we initially called "width." That's fine because these names make no difference in the final calculation of area or volume.

The game ends when any player's building exceeds forty cubes. Finish that round, so all players have the same number of turns. Then count up your score based on the Architect's Prizes below, and the highest score wins.

Architect's Prizes

Award one point in each of the following prize categories:

- Forty or more cubes
- Tallest building
- Greatest volume
- Greatest perimeter around the base
- Most area on any one side

If two or more players tie for an award, each player gets a point.

History

John Golden shared this game on his Math Hombre blog. Teacher Fawn Nguyen created a great follow-up activity that allows more creative freedom for the student architects. Check out Hotel Snap on her Finding Ways blog.†

---

† *fawnnguyen.com/hotel-snap*

Four Corners

MATH CONCEPTS: ordered pairs, coordinate graphing (first quadrant), right angles, parallel and perpendicular lines.
PLAYERS: two or three players/teams.
EQUIPMENT: square grid paper, four six-sided dice, two different colored pencils or markers.

Set-Up

On the grid paper, mark a 12 × 12 square to create the first quadrant of a coordinate graph. Label the x and y axes from zero to twelve.

Or use a page of coordinate graph paper from the free *Prealgebra & Geometry Printables* file.

Players each need a different colored pencil or marker, and they may want to choose a symbol like X, O, or a small triangle or square to make their marks perfectly distinct.

How to Play

On your turn, roll all four dice. You must use all four of these numbers—as rolled, or combined using the standard arithmetic operations—to create the two coordinates of an ordered pair. Mark that point with your colored pencil.

For example, if you roll 2, 3, 3, 6, you could mark:

$$(2, 3+3+6)$$
$$(3-3, 6÷2)$$
etc.

The first player to mark the four corners of a rectangle wins. Yes, squares are rectangles. Tilted shapes count, too, if you can prove they have four right angles.

The losing player goes first in the next game.

Words to Know

An *ordered pair* is any set of two numbers recorded in a particular, defined order. In geometry, we label graph paper with a horizontal number line called the *x-axis* and a vertical number line called the *y-axis*. [The plural of axis is *axes*, with a long e.] Then we describe the position of a grid point by listing the horizontal distance x and the vertical distance y in alphabetical order: (x,y).

We can also call this (x,y) pair the *coordinates* of that point, so the overall structure is called a *coordinate graph*. The section of the grid where both x and y are positive numbers is called the *first quadrant*. That's a natural place for students to make their initial foray into coordinate graphing.

The idea that we can identify position by coordinates is often credited to René Descartes, so this type of (x,y) graphing is sometimes called *Cartesian coordinates*.

Four-sided polygons are *quadrilaterals*. If both pairs of opposite sides are parallel, we call them *parallelograms*. When a parallelogram has all four sides equal, that's a *rhombus*. If a parallelogram has right angles, it's called a *rectangle*. And when a rectangle has all four sides equal, that's a *square*. That means squares are a subset of rectangles (and of rhombuses), and rectangles are a subset of parallelograms.

Variation

SKEWED CORNERS: The first player to create a rectangle or square loses the game.

History

John Golden created this game and shared the story on his Math Hombre blog. He writes:

> "Students made connections to other games, especially Minecraft. (Engagement +2 immediately. 'This stuff is in Minecraft?') Some people felt this was the best game all year. I was impressed how few students were doing the x–y reversal by the end of the game."

Hidden Hexagon

MATH CONCEPTS: coordinate graphing (first quadrant), simple linear equations, irregular polygons.
PLAYERS: two players or two teams.
EQUIPMENT: printed gameboard or square grid paper for each player, pencils, ruler or other straightedge.

Set-Up

To make your own gameboards, give each player a sheet of square grid paper. Players outline two 10 × 10 grids and label the x and y axes 0–10. Label one grid "Top Secret" and the other grid "Clues."

On your secret grid, draw a closed shape with six straight sides, with vertices on the grid points and at least one grid point entirely inside the shape. Use a ruler or straightedge for precise lines. Don't let the sides of your shape cross each other. This is your hidden hexagon. Record the ordered pairs for each vertex beside the grid.

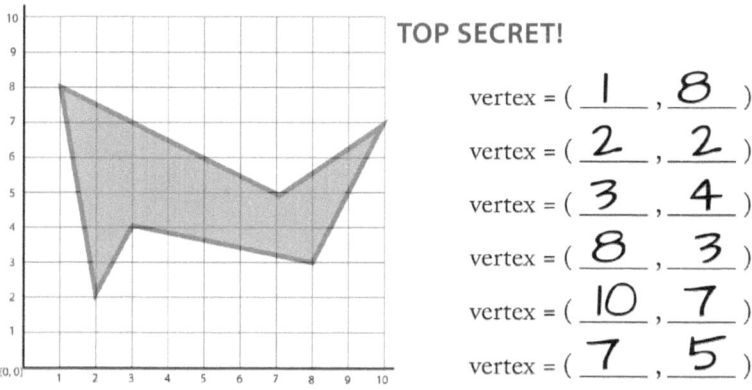

TOP SECRET!

vertex = (1 , 8)
vertex = (2 , 2)
vertex = (3 , 4)
vertex = (8 , 3)
vertex = (10 , 7)
vertex = (7 , 5)

(Fold the paper to hide your secret hexagon.)

You can make your own graphs or use the gameboard from the free *Prealgebra & Geometry Printables* file.

How to Play

Players alternate turns asking for information so they can guess their opponent's secret shape. On your turn, name an equation for a horizontal, vertical, or diagonal line.

For example:

- The equation "$x = 3$" represents the vertical line that includes such points as (3,0) and (3,5).
- The equation "$y = 10$" indicates the top horizontal line on the gameboard, including points like (1,10) and (7,10).
- The equation "$y = x$" is the diagonal line from the origin slanting up at a 45° angle. It passes through such points as (2,2) and (9,9).

Your opponent checks that line against their hidden hexagon and tells whether each coordinate point is a vertex, on the perimeter, inside the shape, or outside.

For example, imagine your opponent has the hexagon shown, and you ask for the $y = 3$ clues. Point (2,3) is inside the shape, and (8,3) is a vertex. All the other points are outside the hexagon.

Record this information on your Clues grid. Invent your own code, such as:

- Vertex = Mark the point with a star and record the coordinates as an ordered pair beside the grid.
- Perimeter = Mark the point with a shaded circle.
- Inside = Draw an open circle.
- Outside = Mark with an X.

After you discover all six vertices of the hidden hexagon, you may guess right away. Or you may keep playing to collect more information on the secret shape.

When you're ready, connect the vertices on your Clues grid. Make

Words to Know

A *hexagon* is a six-sided polygon. It also has six *vertices* (singular *vertex)*, which is the math word for corner points. Is it a coincidence that the number of sides and corners match?

Most people know the familiar honeycomb pattern of regular hexagons. But your children may be surprised to hear that *any* six-sided polygon is a hexagon.

sure the shape fits properly around all the marked points. Reveal your drawing and compare it to the hexagon on your opponent's gameboard.

The first player to guess ends the game. If your drawing matches the hidden hexagon, you win the game. But if your guess is wrong on any point, the other player wins.

Variation

You can play Battleship-style: Players name individual points instead of a full line. If the game seems too long, try asking for two or three points per turn.

Exploration

After playing a few rounds, take time to compare observations. Think about the different shapes you've seen in your games. What tricks make a hidden hexagon difficult to guess?

Can you think of any other ways to draw a six-sided shape? Try making a concave hexagon, or one with no symmetry at all. Can a hexagon have a right angle? Can it have a *reflex* angle, greater than 180°?

If you use a whole page of square grid paper, does that give you more options for weirdly shaped hexagons? What if the vertices don't have to be on a grid point?

Can you make a conjecture about hexagons?

History

Teacher coach Terry Kawas modified the traditional Battleship game to create a quadrilateral guessing game called Shape Capture. Teacher and author Christopher Danielson shared a wonderful geometry investigation called Hierarchy of Hexagons. If ideas can have children, this game of mine is their offspring.

Words to Know

On a coordinate graph, the horizontal *x*-axis number line and the vertical *y*-axis number line meet at the point (0,0), called the *origin*. The origin is the point from which we measure all other coordinates.

Like zero on a number line, the origin is an arbitrary point. It can go anywhere on your paper.

The horizontal and vertical axes divide the coordinate plane into four areas called *quadrants*. In the first quadrant, both *x* and *y* coordinates are positive numbers.

If we draw the axes in the traditional position, so that the first quadrant is in the top right-hand section of the page, the other quadrants are numbered in order counterclockwise. What types of (*x,y*) numbers do you find in each quadrant?

Coordinate Gomoku

MATH CONCEPTS: ordered pairs, coordinate graphing (four quadrants).
PLAYERS: two players or two teams.
EQUIPMENT: dotty or lined square grid paper, different colored pencils or markers.

Set-Up

Players share one sheet of dotty or lined grid paper. Each player needs a different colored pencil or marker, and they may want to choose a symbol like X, O, star, or small triangle to make their marks perfectly distinct.

How to Play

The first player claims any dot on the grid by marking it with his or her symbol. If you are using lined graph paper, choose any place where two lines intersect.

This position is the *origin* (0,0) for your game. Draw the horizontal and vertical lines (the x and y axes) that meet at that point.

On succeeding turns, you must name the coordinates of a point before you mark it. If you make a mistake naming the point, your opponent can give the correct (x,y) coordinates and mark it with their own symbol—and they still get to take their own turn.

The first player or team to mark five points in a horizontal, vertical, or diagonal row with no gaps wins the game.

Variations

HOUSE RULE: How do you want to handle *overlines* that have six or more points in a row? In traditional Gomoku, only an exact five-in-a-row line can win.

Swap2: Does the first player win too often? What you need is a variation on the "pie rule"—one person slices the pie, and the other gets first choice of piece. Reduce the first-player advantage with these starting moves, which are often used in tournament play. The first player marks two Xs and one O. The second player selects one of three options:

- Accept these moves and mark an O. Turns continue with the first player marking X.
- Take over the Xs, so the first player goes next with O.
- Or mark one additional X *and* O, then let the first player choose which letter to claim. Whoever plays O takes the next turn.

History

Gomoku is a Japanese game played with stones on the Go board. Math teachers have played coordinate-graph Gomoku games with students for decades. The Nrich Maths website added the idea of letting the origin float to wherever the first point is marked.

Linear War

MATH CONCEPTS: coordinate graphing, linear equations, slope, intercepts.
PLAYERS: two to four.
EQUIPMENT: printed or homemade game cards, pencils or markers, ruler or other straightedge.

Set-Up

Players each create a personal deck of eleven game cards. The printable download file includes coordinate grid templates for the game cards. Print on card stock or sturdy paper for best results. Or cut regular grid paper into 10 × 10 squares.

Draw coordinate axes with the origin at the center of each card. Number the points from −5 to +5 on each axis.

On each card, draw a line that passes through at least two points with integer coordinates, such as (−3,1) and (0,5). Mark these points with dots on the line.

Claim your deck: Mark the back of each card with your insignia (initials, emoticon, math symbol, etc.—your choice). Decorate as desired, so long as it doesn't bleed through the paper and obscure your lines.

How to Play

Place your game card deck face down in front of you and draw the top two cards for your hand. The player whose turn it is names the trump:

- ♦ Least slope or greatest slope
- ♦ Least or greatest x-intercept
- ♦ Least or greatest y-intercept

Words to Know

An *intercept* is the point where a line meets either the x- or y-axis. Because the axes cross at the origin, they each mark the points where the other coordinate is zero. So for your line, the x-intercept is the value when $y = 0$. And the y-intercept is the value when $x = 0$.

The *slope* is the steepness of a line when you read across the graph from left to right, as in reading a book. A line that slants upward has a *positive* slope, and a line running downhill has a *negative* slope.

The slope is also called the *rate of change* for your line. It tells how much the y value increases with every one-unit increase in the value of x. If the slope is negative, that means y decreases as you increase x.

If two cards have slopes that look almost the same, you may need to calculate their value to decide the winner. Use the two dotted points (with integer coordinates) as follows:

- Count the change in y value from one point to the other. If the number goes down, that is a negative change.
- Count the change in x value from the same point to the other. Always measure these changes in the same direction.
- Divide the change in the y coordinates by the change in x. That quotient is the slope of your line.

For a flat (horizontal) line, the y coordinate never changes, so you have zero divided by the change in x values. Flat lines have *zero slope*. Can you explain why vertical lines have an *undefined slope*?

Each player chooses a card from their hand and holds it face down. When all players are ready, turn the cards up, and the card that best fits the trump captures them all. Each player keeps a pile of prisoner cards.

In case of a tie, leave the cards on the table for the winner of the next trick.

Players draw a new card to replenish their hands. When the decks run out, continue playing until all the cards are taken or the last round ends in a tie. Whoever captures the most prisoners wins the game.

Give the prisoner cards back to their owner, unless you are playing the Spoils of War variation.

Variations

In between games, players can make new cards. But they must always choose a deck of eleven cards for play.

HOUSE RULE 1: Are players allowed to draw lines with undefined slopes? If so, how will you treat those cards when the trump is "greatest slope"?

HOUSE RULE 2: How will you handle cards whose x- or y-intercept is off the chart? Are they allowed to compete for the intercept trumps categories?

SPOILS OF WAR: If you win the game, you may claim one card from your prisoner pile to replace a card in your deck. Erase the identification and add your own symbol on the back. Offer your old card to the losing player as a consolation prize, if you wish.

History

This is yet another game from the inventive mind of John Golden, the Math Hombre. He suggests using it throughout a unit on linear graphing, having students make different game cards as you introduce each vocabulary term. At the end of the unit of study, they can choose their eleven-card decks and play the game for review.

Radar

MATH CONCEPTS: angle measures, polar coordinates.
PLAYERS: two to four.
EQUIPMENT: printed gameboard or polar (circular) graph paper, pencils or markers.

Set-Up

Choose a gameboard from the *Prealgebra & Geometry Printables* file or make your own on polar graph paper. To create your own gameboard, mark circles of radius 1, 2, 3, etc. The number of circles must be equal to or greater than the number of people who want to play.

Around each circle, draw lines at convenient angles such as 30°, 60°, 90°, and so on. Or use radians: π/6, π/3, π/2, and so on.

Finally, draw an X or poison symbol in the part of the graph farthest from the origin, just below the θ = 0 axis. Whoever is forced to take that section of the graph loses the game.

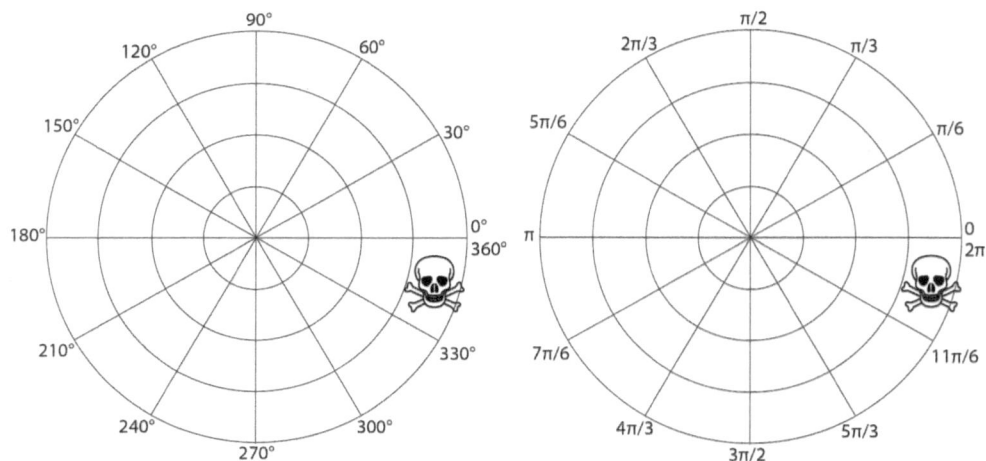

Each Radar gameboard has room to play four games.
You can move to any point where an angle line intersects a circle.

How to Play

Imagine each move as a radar detector sweeping around the graph counterclockwise from the θ = 0 axis. The first player chooses any intersection point on the graph, stating the move in polar coordinates (*r*,θ). Shade in the wedge-shaped sector swept out by that move.

On succeeding turns, players must increase either the radius or the angle, or both. Shade in the additional sections covered by each move.

A player who increases the angle may decrease the radius, if desired. Or a player who increases the radius may decrease the angle. But the radius can never go below *r* = 1 and the angle can never go less than θ = 30° or $\pi/6$. (Or the smallest angle marked on your circle.)

Players must shade in at least one section of the graph on every turn.

The player forced to color the last section of the graph, completing the circle at the maximum radius, loses the game.

Sample Game

Fezzik challenges Inigo to a game of Radar. They choose a gameboard with radius 4 and angles measured in radians.

Fezzik moves first, claiming the point (2,$\pi/3$). He colors in the wedge from that point to the origin and down toward the θ = 0 starting axis.

Inigo takes (3,$\pi/6$), increasing the radius but going for a smaller angle. He colors his sector down to the axis and inward until it meets the already-shaded area.

Fezzik chooses (1,$5\pi/6$), almost halfway around the smallest circle. He shades the wedge between that point and his previous move.

Inigo gets impatient and takes a big chunk. He marks (4,$\pi/2$) and colors in the rest of the first quadrant.

The shaded area will continue growing until one player has no choice but to mark the point (4,2π) and take the poison sector.

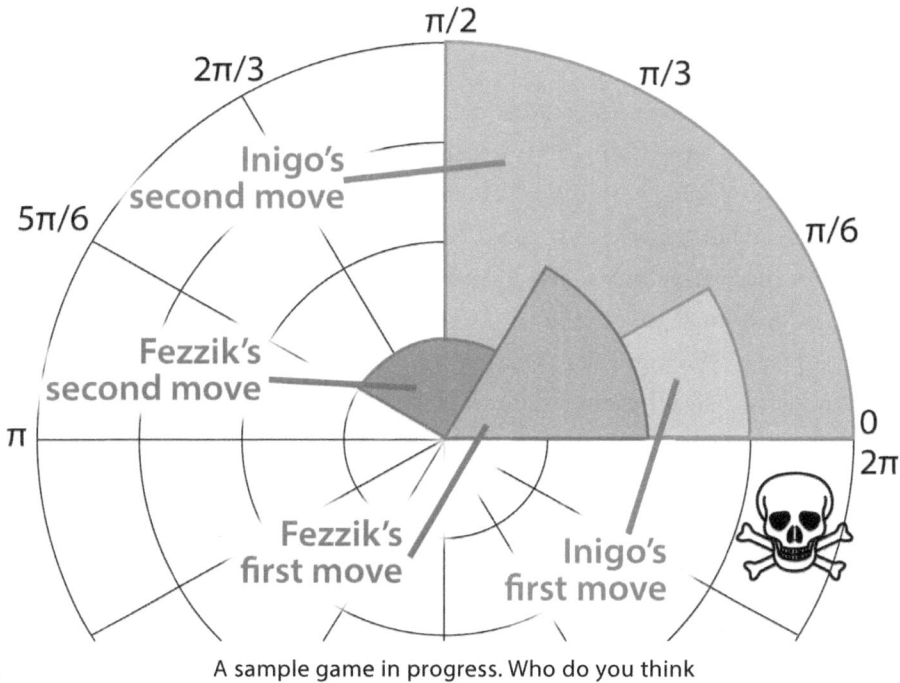

A sample game in progress. Who do you think will get stuck with the poison?

Variations

RADIAN RACE: Each player chooses one circle and places a small game piece at $\theta = 0$. On your turn, slide your piece around your own circle to the angle of the farthest player's position, and then move your choice of one or two angle sections further. The first player to make a full trip around the circle, arriving back at the starting point, wins the game.

RADIAN NIM: Play as in the race above, but as a *misère* game. The first player to complete the circle loses.

History

Radar is my circular version of the two-player strategy game Chomp, which inventor David Gale described as a curious way to eat a chocolate bar. Author Martin Gardner popularized the game in his *Scientific American* column "Mathematical Games."

Words to Know

We can use *polar coordinates* to name any point on a graph by its distance from the origin *r* and the counterclockwise angle θ *(theta)* from the positive *x*-axis. Except in a polar graph, it's no longer the *x*-axis; it's the θ = 0 axis.

We can measure angles in degrees or radians. Degrees are an arbitrary unit chosen by astronomers in ancient Mesopotamia or Persia for easy calculation, because 360 has many factors. But radians are deeply connected to the mathematics of a circle.

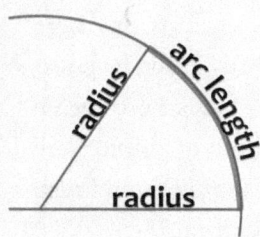

A *radian* is the angle that sweeps out an *equilateral* (equal-sided) *wedge*—that is, a slice-of-pizza shape where the arc length of the circle exactly matches the radius.

We usually measure radians in terms of *pi* (π), which is the number that relates length-around to length-across. The *circumference (C)* is the full distance around a circle, and the distance across the center of a circle is the *diameter (d)*. Pi measures the number of diameters it takes to make a circumference:

$$\pi = C \div d = C/d$$
$$= 3.1415926535...$$

All circles are the same basic shape, only enlarged or reduced. So π is the same number for every circle. It takes a little over three times across the circle to make the same distance as going all the way around.

Since the *radius* is half the diameter, there are π radians in a half-circle:

$$\pi \text{ radians} = 180°$$

And the circumference is 2π times the length of the radius. So there are 2π radians in a full circle.

Racetrack

MATH CONCEPTS: vectors, coordinate graphing, delta notation for changing coordinates.
PLAYERS: two or three.
EQUIPMENT: dotty or lined square grid paper, pencils or colored markers, ruler or other straightedge.

Set-Up

Make your own racetrack on grid paper by drawing an elliptical or irregular path. The track may have wide and narrow sections, but there must be at least two squares open between the walls at every point. Before the race starts, players should examine the track and decide whether any questionable points are on the road or out of bounds.

Draw two straight lines across the track to mark the start and finish lines for your race, or one line on a track that loops around.

Also make a chart with a column for each player. Players choose a color or symbol (such as X and O) to represent their cars. Write the name and symbol at the top of each player's column.

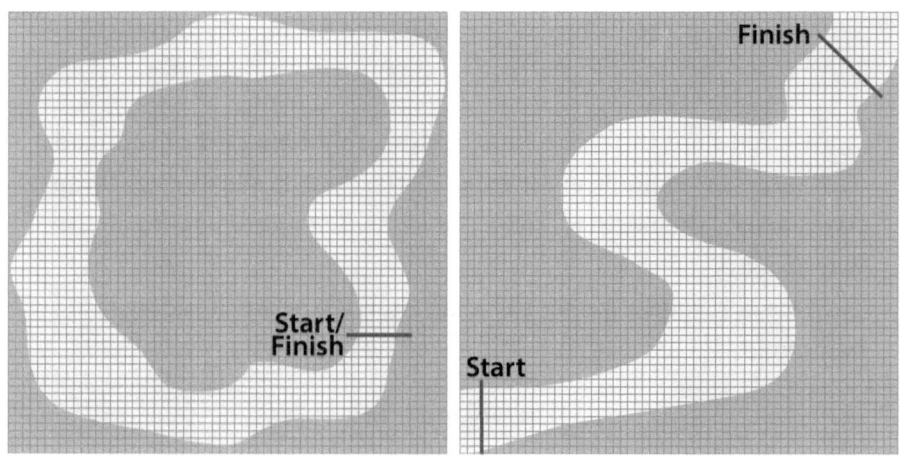

Two sample racetracks. You can race on a
one-way track or around a loop.

180 ♦ PREALGEBRA AND GEOMETRY GAMES

How to Play

Each player marks a dot or grid intersection on or behind the starting line. Players move in the opposite order of how they chose their starting position—that is, whoever got first choice of position moves last.

On each turn, you travel along a vector ($\Delta x, \Delta y$). This means you make some change in your horizontal position x, and some change in your vertical position y, measured relative to where you started that turn.

For example, a vector of (−1,2) means:

- Move −1 in the x direction, one square to the left.
- Move +2 in the y direction, two squares up on your grid paper.

Write the movement vector in your column on the gameboard. Then put a symbol at your new position and draw a straight line connecting it to your previous spot.

For your first turn, move one square in any direction. That is, you must choose the values for your initial movement vector from this list: −1, 1, or 0, mix or match. After that, you may change the Δx value, the Δy value, or both, by one square per turn. Or you can keep cruising at the same speed.

So after the (−1,2) move described above, you might:

- Coast along with another (−1,2) movement.
- Speed up to (−2,3), two squares left and three squares up.
- Slow down to (0,1), no horizontal movement and only one square up.
- Or any combination of these changes.

The lines that mark each player's path may cross, but a player may not move directly through another car's current position. Also, players must stay within the boundaries of the track for the entire length of their moves, or else they crash and lose the game.

Words to Know

A *vector* is a force or movement that has both magnitude and direction. Back in Chapter 4, I mentioned that we can think of integers as vectors.

When we write polar coordinates, we are writing vectors. The coordinates (r,θ) represent that point's *(magnitude, direction)* from the origin point (0,0).

We can also use vectors on an *x-y* Cartesian coordinate grid. In that case, we don't write the magnitude and direction precisely, but instead show the relative change in position. We can think of the standard coordinates (x,y) as representing the change in position from the origin (0,0) to that point.

In the Racetrack game, we represent a movement vector (also called a *translation*) by writing an ordered pair $(\Delta x, \Delta y)$. Read this as "delta-*x*, delta-*y*." The Greek capital delta (Δ) means "the change in." So Δx is the change in horizontal *x* position. And Δy is the change in vertical *y* position.

Many people write translation vectors vertically $\begin{pmatrix}\Delta x\\ \Delta y\end{pmatrix}$ to distinguish them from ordinary coordinates. The numbers sit one above the other, like a fraction without the bar. You may use either method to record players' movements in the Racetrack game.

You can learn more about vectors and how to convert between polar and Cartesian coordinates—and many other topics—on Rod Pierce's Math Is Fun website.[†]

[†] *mathsisfun.com/algebra/vectors.html*

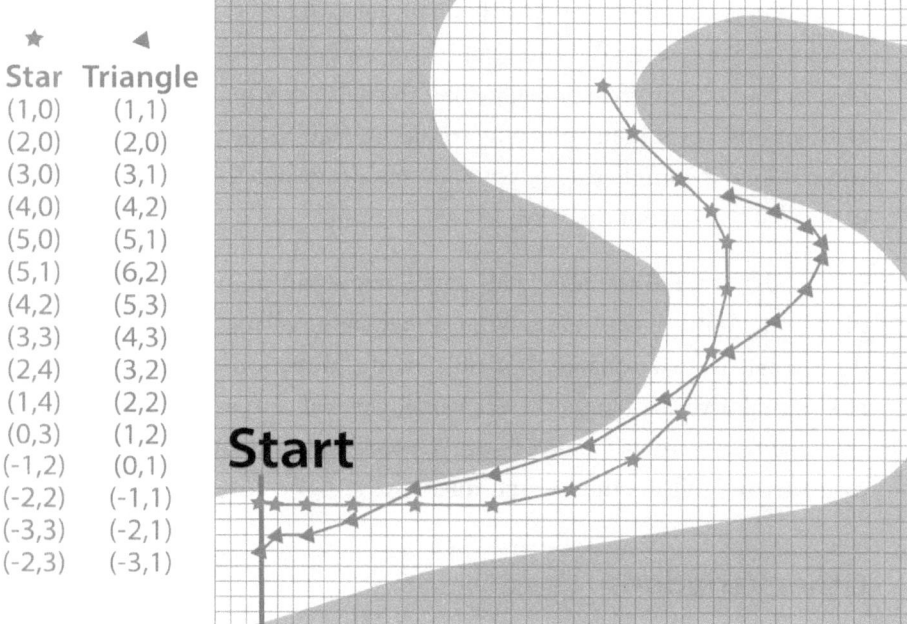

★ Star	◄ Triangle
(1,0)	(1,1)
(2,0)	(2,0)
(3,0)	(3,1)
(4,0)	(4,2)
(5,0)	(5,1)
(5,1)	(6,2)
(4,2)	(5,3)
(3,3)	(4,3)
(2,4)	(3,2)
(1,4)	(2,2)
(0,3)	(1,2)
(-1,2)	(0,1)
(-2,2)	(-1,1)
(-3,3)	(-2,1)
(-2,3)	(-3,1)

The beginning of a game of Racetrack, showing the movement vectors and each car's path. The triangle player jumped ahead at first but went too fast and almost didn't make the turn.

The first person to travel around the track and cross the finish line wins the race. But any players who have not yet moved during that turn may complete their moves. If more than one person crosses the finish line in the same turn, then whoever goes the farthest past it is the winner.

Race Hazards (Optional)

Before the race starts, players may lightly color sections of the track to mark them as hazards. For instance, you might try…

OIL SPILLS: Shade light gray with pencil. Cars traveling through an oil spill cannot get traction to accelerate. If you pass through this area, your movement vector must repeat your previous turn.

LOOSE GRAVEL: Fill the area with small dots. Gravel slows you down and limits your control. When you drive on gravel, both coordinates of your movement vector must be half the value of your previous turn. With an odd coordinate, you may round the half-value up or down. If you come to a complete stop, you may move one square on your next turn, as at the start of the race.

TURBO BOOST: Fill the area with small arrows all pointing the same direction. Within this area, your movement vector increases (or decreases) by two squares, but only in the direction of the arrows. Horizontal arrows change the Δx value, vertical arrows the Δy value, and diagonal arrows change both. A coordinate not affected by the turbo arrows follows the normal rule.

Other Variations

HOUSE RULE: Do you hate to lose the game because of a crash? Choose a rule that gives players a chance to recover from miscalculations.

For example, players who drive off the track can move one square per turn until they return to the racetrack.

Or the player moves one square per turn, and the car must reenter the track behind the place where it went off.

Or players may return magically to any point behind where they left the track. They lose their next turn—a (0,0) movement vector. After that, they accelerate as at the beginning of the game.

History

Racetrack is also called Vector Race or Graph Paper Race. In 1973, Martin Gardner wrote about the game for his "Mathematical Games" column in *Scientific American*. Michael Serra explained how to use vector notation for player moves, Dan Meyer blogged about it, and I saw it on Meyer's dy/dan blog.

Section III

Playing to Learn Math

Math develops reasoning skills — I hear it all the time. But math only develops reasoning skills if we allow kids to reason.

— Sonya Post

More Than One Way to Solve It

SOME PEOPLE TELL ME THEY like math because of its certainty. If they follow the proper steps, they'll get the right answer. The goal is clear, unlike a language essay question. There's no place for bias or difference of opinion, unlike trying to interpret historical cause and effect.

I understand the sentiment. But that's not what I like about math.

To me, the attraction of math is in the problem-solving journey itself. There may be only one right answer to a math problem. But there is always more than one way to find that answer.

I love the adventure of exploring those different ways of thinking.

Let me show you what I mean. Follow me on a memory trip down one of my favorite middle-school rabbit trails: learning how to count.

Our trek began the day my youngest daughter complained about her math workbook.

"It's all the same thing," she said. "Repeating over and over, just with bigger numbers. The problems never get any harder, they just get longer. It's boring!"

So we put that book away and sat together on the living room couch, sharing a whiteboard on our laps. We spent the next several months working through the counting and probability lessons in Jason Batterson's *Competition Math for Middle School*. My daughter has no interest in competing, but she enjoys learning new ideas.

The Science of Counting

New Idea #1: Not Only for Little Kids

Counting is part of *discrete mathematics,* which means the study of things that are not continuous—things that come in discrete chunks—and it's important in game design, computer programing, and cryptography.

We spent a bit of time counting basic numbers and counting items in sets. Then we moved to the more interesting type of mathematical counting. *Combinatorics* is the art of counting possibilities: all the choices we can make, the things we may do, the ways we might combine stuff.

For example, imagine you are choosing your avatar for a new online game. You need to pick an outfit, and you have the following options:

- Leggings: brown or black, tucked into high leather boots.
- Top: green, purple, or blue.
- Cloak or no cloak.
- Head piece: hat, scarf, eye patch, or none.

Focus on the first two choices, the leggings and top. Imagine these choices as paths through a mental landscape, branching off each other to form a *tree diagram.* Your first option creates two paths: brown leggings, or black leggings. Each of these paths leads to a new question: Which top? At that point, you must choose between three branching paths.

The total number of branches in a tree diagram is the product of the choices at each branch point. In this case, 2 × 3 = 6 possible combinations.

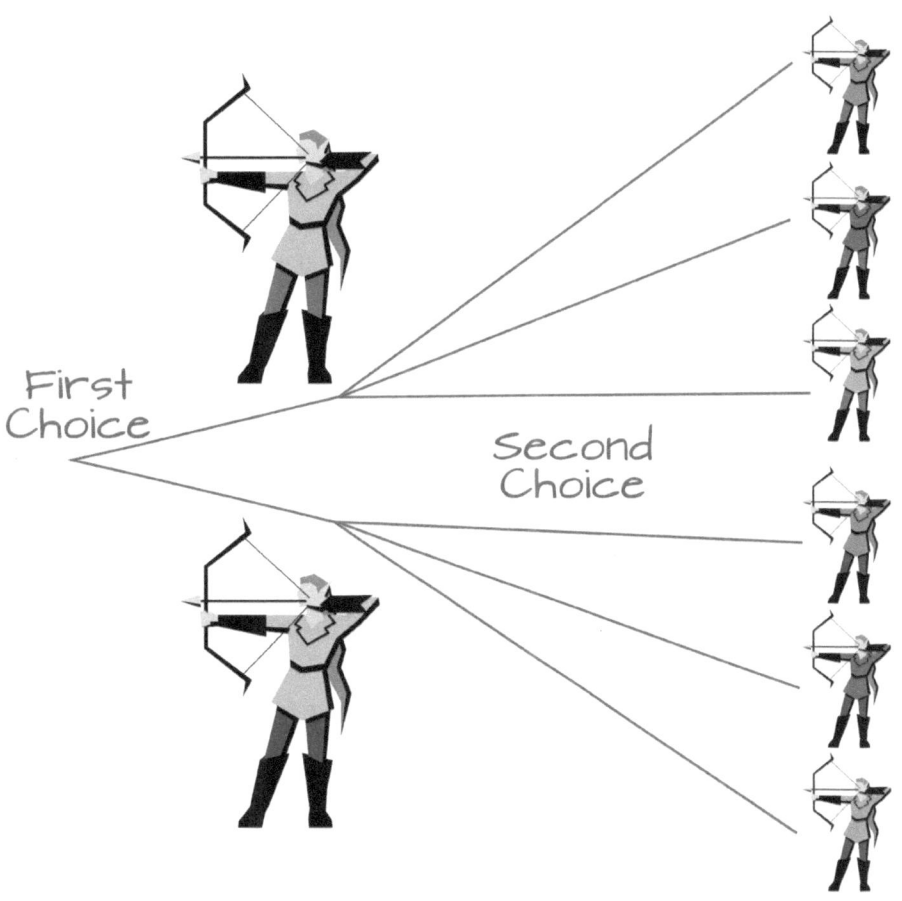

You can visualize simple choices as the branching paths of a tree diagram.

New Idea #2: The Fundamental Counting Principle

As we add more choices, the paths continue to branch out. Our tree diagram grows unwieldy, and we desperately need a new counting tool.

The *fundamental counting principle* comes to our rescue. We can count the possibilities of a situation by multiplying our options.

> *If you have A options for your first choice, B options for your second choice, C options for your third choice, and so on,*

then you can find the total number of potential outcomes by multiplying:

$$A \times B \times C \times \ldots$$

To choose our game avatar, we have to make four choices—leggings, top, cloak (or not), and head piece—with two, three, two, and four options respectively. So we calculate the total number of possible outfits:

$$2 \times 3 \times 2 \times 4 = 48$$

It all made sense to my daughter. We whipped through the problems. Until…

A Pizza Conundrum

Pizzas at Mario's come in three sizes, and you have your choice of ten toppings to add to the pizza. You may order a pizza with any number of toppings (up to ten), including zero. How many choices of pizza are there at Mario's?

Conditioned by the easy problems earlier in the lesson, my daughter mentally multiplied the number of crust options times the number of toppings, for $3 \times 10 = 30$ different pizzas.

If I had been thinking well myself, I would have encouraged her. That would be correct *if* we wanted the number of one-topping pizzas.

As soon as I said her answer was wrong, she realized there had to be more options.

She started thinking about possibilities, writing the calculation on her whiteboard. three choices for the crust size, times ten choices for the first topping, times nine choices for the second topping, times…

Her instinct was right, but the counting principle is tricky in this problem. The actual choice for each topping is a simple "yes" or "no." So the numbers she needed were:

$$3 \times 2 \times 2 \times 2 \ldots$$

Wanting to save her from the frustration of a nearly endless calculation that I knew would end in another wrong answer (and possibly induce an emotional melt-down), I spoke up. "Sometimes, when a problem is too tough, a good thing to try is to make it simpler. What could—"

She dropped her marker and hunched down over the whiteboard with eyes clenched tight, both hands over her ears.

I took the hint and shut up.

We sat there.

And sat.

And.

Sat.

After what seemed like forever (but was probably less than five minutes), she picked up her marker and made a list:

$$
\begin{array}{c}
C \\
P \\
C\text{-}P \\
P\text{-}C \\
N
\end{array}
$$

"With two toppings, we get five pizzas," she said.

Ah, her code meant C for extra cheese, P for pepperoni. N must be "no toppings" for a plain cheese pizza.

I pointed out that most pizza places have their own rules about which order the toppings go on, so they wouldn't serve both Cheese-Pepperoni and Pepperoni-Cheese.

She agreed. "The cheese should always go on first, under the other toppings. That way you can pick off whatever you don't like."

With a swipe of the eraser, two toppings gave her four pizza options.

Her next topping was ham (H), and she had no trouble finding the eight combinations for an up-to-three-topping pizza.

We were momentarily stumped on a fourth topping. Peppers or pineapple wouldn't work, since she already had a P, and mushrooms were unthinkable. She decided to use O. (Onion? Olives? Olives would definitely get picked off.) It was trickier to make sure we had all the possibilities, but in the end she listed sixteen combinations.

"Each new topping makes it twice as many," she said. "So there are 3×2^{10}."

$$3 \times 2^{10} = 3{,}072 \text{ different pizzas}$$

That seemed reasonable. Every time you put a new topping on the menu, you can still order all the options you had before. Or you could make each of those choices with the new topping added. Double your pleasure.

Another Day, Another Puzzle

In a lazy I-don't-want-to-do-school mood, my daughter was ready to stop after three math problems. We got two of them correct, but the question of how many ways we might paint the sides of a cube in black and white stumped us.

For my young perfectionist, one mistake was excuse enough to quit. She sighed and leaned her head against my shoulder as we sat together on the couch.

"We're done," she said. "Done, done, done."

Still, I squeezed in one more puzzle. I picked up my whiteboard marker and started writing:

>DONE
>DOEN
>DNOE
>DENO
>DNEO
>ODNE
>ODEN

We were still working our way through the lessons in *Competition Math*. We'd finished Venn diagrams, triangular numbers, and the fundamental counting principle. That day, we studied *casework,* which means counting the various possibilities for each case.

The cube puzzle, for instance, had four cases: solid color, one side different, two sides different, and half-and-half. We had to consider the number of possibilities under each case, but we focused too much on how to orient the different-colored sides. We forgot to count the two solid-color options.

But I figured this "DONE" problem was review, a fairly basic application of the counting principle. Four choices for the first letter, three options for the second, two remaining for the third, and only one letter left at the end.

$$4 \times 3 \times 2 \times 1 = 24$$

Naturally, that's not the way my daughter figured it out.

Why does it still surprise me when she thinks differently than I do? I should expect it by now.

She pointed at my list.

"If we have five ways to start with D," she said, "then we could switch the D and the O. So then we'd have all the same ways to write it, except with the D replacing the O and the O replacing the D.

"Or we could switch the D and the E, or the D and the N.

"That means we really have four times as many ways. It's simple!"

That made sense. It should work.

But we had a problem. Using the fundamental counting principle, I had calculated twenty-four ways to arrange the letters. My daughter found twenty solutions with her "letter switching principle." But no matter how we count them, the number of arrangements shouldn't change.

The answers have to match.

At least one of us had made a mistake.

I looked back over my list. Sure enough, I had missed one way to start with D. I added it to my list:

DONE

DOEN

DNOE
DENO
DNEO
and one more: DEON

Now her letter switching principle gave us 6 × 4 = 24 arrangements. Success!

Counting with Complications

Two months later, we were still having fun with our counting lessons—and my daughter continued to demonstrate that she doesn't think the same way I do.

But this was one of those days when we spent a half-hour on math and barely finished a single problem:

How many of the possible distinct arrangements of the numbers 1–6 have 1 to the left of 2?

We each used a separate lap-sized whiteboard. Puzzling things through for myself kept me from talking too much. While I worked, my daughter had time to think.

For this problem, I tried casework—that is, I considered where the 1 might go and still have 2 on its right. Math problems can be tricky because language is imprecise. But the problem didn't say *"directly to the left,"* so I figured other digits could come between them. That gave me five cases to count:

```
1 _ _ _ _ _
_ 1 _ _ _ _
_ _ 1 _ _ _
_ _ _ 1 _ _
_ _ _ _ 1 2
```

For every case, I filled in the number of choices available for each blank space, taking into account the rule about the digit 2. Using the fundamental counting principle, I multiplied everything out (five multiplications, one for each line). And finally, I added those products together to find my answer.

Casework is like hitting the problem with a sledgehammer. It's messy, but it gets the job done.

My daughter was still thinking, doodling numbers on her whiteboard and then erasing them.

To keep from breaking her concentration, I decided to compare my result with the answer in the back of the book. The author used symmetry to solve the problem, recognizing that every six-digit number, such as:

$$1\ 2\ 3\ 4\ 5\ 6$$

... has a mirror-image partner:

$$6\ 5\ 4\ 3\ 2\ 1$$

And in every mirror-image pair, only one number has the 1 and 2 in the correct order. So we can count the total number of ways to arrange the digits and then divide by two.

Most people think of symmetry as a geometric concept, but it can be a useful problem-solving tool in many areas of math.

Build the Number, Counting As You Go

My daughter was ready to share her answer. Her method took only two short lines on the whiteboard, accompanied by her oral proof.

To begin with, she explained, the 1 and 2 must be in this order:

$$1\ 2$$

This is the only option for those two digits, so she counted the total number of potential six-digit numbers using the fundamental counting principle:

$$1 \times [\text{the choices for the other digits}]$$

I almost interrupted her to say that the numbers don't have to sit next to each other. Thankfully, I kept my mouth shut, and thus saved myself from sticking my foot in it. She had everything under control.

The next digit could fit before the 1, or after the 2, or in between them:

$$\_\,1\,\_\,2\,\_$$

That digit had three options, making the total so far:

$$1 \times 3 \times \text{[the choices for the other digits]}$$

She added 3 at the end of the line, for demonstration. But she reminded me that the possibilities were the same no matter which digit we used or where we put it.

The next digit had four options of where to go:

$$\_\,1\,\_\,2\,\_\,3\,\_$$
$$1 \times 3 \times 4 \times \text{[the choices for the other digits]}$$

She wrote 4 at the end of her growing number.

The next digit had five options of where to go:

$$\_\,1\,\_\,2\,\_\,3\,\_\,4\,\_$$
$$1 \times 3 \times 4 \times 5 \times \text{[the choice for the last digit]}$$

The final digit had six options of where to go:

$$\_\,1\,\_\,2\,\_\,3\,\_\,4\,\_\,5\,\_$$

Final count of six-digit numbers that fit our criterion:

$$1 \times 3 \times 4 \times 5 \times 6 = 360$$

I was impressed. Her solution felt so elegant that I think she could have a future as a mathematician. After all, every aspiring novelist needs a day job, right?

If only I could get her to give up the idea that she hates math.

The Moral of the Story

"A good problem should be more than a mere exercise; it should be challenging and not too easily solved by the student, and it should require some 'dreaming' time."
—HOWARD EVES

The point of these stories is not to brag on my daughter (proud mama though I am), but to show what mathematical thinking looks like. If we want to teach our children to reason, we must let them struggle through the wilderness of not-knowing.

Our students need to be comfortable with the idea that some answers will be wrong. That paths which look promising may lead nowhere. That it could take more than half an hour to solve a single math problem. That they might get stuck going round and round and never find a solution.

And that's okay.

After all, it took mathematicians over 350 years to find a solution to Fermat's Last Theorem. But all their various failed attempts led

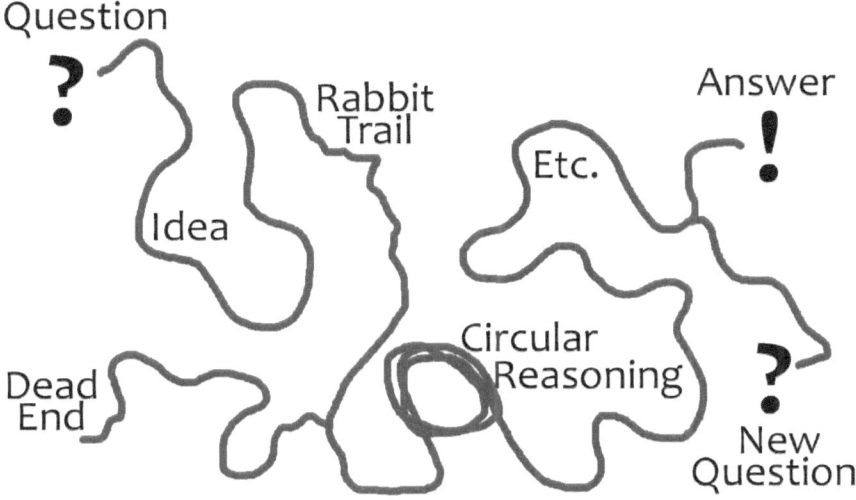

Trying to solve a math problem can feel like forging a trail in the wilderness.

to plenty of new mathematics nobody might ever have discovered otherwise.

We adults want to make life easy for our kids. We're tempted to offer them hints, tricks, methods, and rules that pave a highway from question to answer.

But it's only in the struggle that children learn how to think for themselves.

The teacher who wishes to serve equally all his students, future users and nonusers of mathematics, should teach problem solving so that it is about one-third mathematics and two-thirds common sense.

— GEORGE POLYA

Conclusion: Algebra Tells a Story

IN AMERICAN CULTURE, IT'S FASHIONABLE to denigrate the value of high school math, beginning with algebra.

Not long ago, retired political science professor Andrew Hacker wrote an opinion piece for *The New York Times*. He recommended doing away with algebra class, replacing it with statistics and real-world economic analysis. Hacker cites indisputable problems with school math as currently taught, and the topics he mentioned certainly deserve our students' attention.

The problem is, no one can truly master these valuable citizen-math ideas without a solid foundation of algebra. It's impossible.

As mathematician and popular author Keith Devlin explains, "Hacker and I have very similar views about the abysmal state of much of today's K–12 mathematics education. Unfortunately, he proposes we throw away the healthy but neglected baby along with the depressing pool of lukewarm, dirty bathwater it currently hides in."

This debate about math education isn't new — it dates back at least to the early 1900s. In 1915, the influential education professor William Heard Kilpatrick called algebra "an intellectual luxury" and recommended schools limit themselves to teaching practical mathematics.

Instead of eliminating algebra, what we really need to do is teach algebraic thinking from the very beginning of our children's adventure with math. That way, when they get to high school, the extension into abstract realms won't come as a shock, but as a further application of common sense.

Perhaps it would help if we could all agree on what we're talking about.

Like many people, Kilpatrick and Hacker see algebra as doing math with letters and symbols, following a textbook's capricious rules. Endless vocabulary, complex formulas to memorize, and word problems about Train A leaving from Chicago to meet train B coming in from New York. What value could there be in something so blatantly artificial?

But while that description may to some degree fit the subject taught in school, it's as far from true algebra as paint-by-numbers is from true art.

Like most words, *algebra* has many definitions. Which one applies will depend on context, but none of them involve memorizing arbitrary rules.

Mathematician and educator Mary Everest Boole defined algebra as "dealing logically with the fact of our own ignorance. Algebra deals not with laws of number, but with laws of human thinking."

In Boole's view, algebra means using what you know to figure out something you don't know. It's a mental game of creative problem solving. In that sense, my daughter's foray into discrete mathematics required plenty of algebraic reasoning.

Italian polymath Galileo Galilei wrote, "The universe stands continually open to our gaze. But the book cannot be understood unless one first learns to comprehend the language and read the letters in which it is composed. It is written in the language of mathematics."

Algebra is the language of scientific thought. By condensing huge amounts of data into concise patterns, algebra helps us understand the world. It lets us summarize difficult concepts with one-line equations. With algebra, we can discover relationships and predict how changing part of the problem will affect the final result. It's a tool for wrapping our brains around such big ideas as gravity, statistics, student loan interest, or the spread of a new disease—those important citizen-math topics we all want our children to learn.

Activist and teacher Bob Moses, the author of *Radical Equations*, knows the significance of algebra to every student's future. He calls it a civil right. "Education is still basically Jim Crow as far as the kids who are in the bottom economic strata of the country," he says. "Mathematics joins reading and writing as a literacy needed for full citizenship."

But my favorite definition of algebra comes from Sue VanHattum, editor of the book *Playing with Math* and author of the Math Mama Writes blog, who describes algebra as a mathematical story:

> "We say math is a language. If it is, then we should be able to tell stories in it. Each equation makes a statement, and those statements tell stories.
>
> "I bought a tree and planted it. It was one foot tall at first. It grows two feet a year. Let's make a data table for height versus time, and a graph, and an equation.
>
> "The graph and the equation both tell the story of the tree."

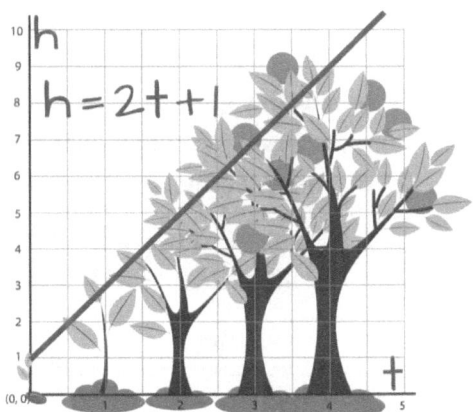

Algebra works with geometry to tell the story of a growing tree.
Each grid square represents one-half year on the
input *(t)* scale and one foot of output *(h)*.

Children can reason through stories. They know how to understand a narrative. To teach our students math, we need to help them see the story behind an equation, a graph, or a formula like the fundamental counting principle.

When students see how the letters and symbols of algebra tell a story, they'll be able to deal logically with their own ignorance. Then we can teach them how to apply common sense to solving math problems.

Online Resources

Now we've reached the end of my *Math You Can Play* series of books. Yet your children's exploration of the mathematical landscape is only beginning.

Where can you go from here? Perhaps you'd like to investigate some of the following online resources.

- ♦ Play around with common-sense algebraic thinking at the SolveMe Puzzles page. And give Paul Salomon's imbalance problems a try, too.[†]
- ♦ Experiment with the "growth stories" algebra can tell on Fawn Nguyen's Visual Patterns website. Or help your students see graphs as stories with Jenna Laib's Slow Reveal Graphs website.[‡]
- ♦ Launch your own adventures in learning to count with the puzzles at Jason Batterson's AGmath site, followed by James Tanton's Permutations and Combinations video course.[§]
- ♦ Discover a wonderful variety of activities on Gordon Hamilton's Math Pickle website. Be sure to visit his

---

[†] *solveme.edc.org*
lostinrecursion.wordpress.com/resources/imbalance-problems
[‡] *visualpatterns.org*
slowrevealgraphs.com
[§] *agmath.com/media//DIR_57727/1$20Counting.pdf*
agmath.com/media//DIR_57727/Probability.pdf
gdaymath.com/courses/permutations-and-combinations

collection of unsolved math problems.[†]

♦ Don't miss the Nrich Maths site, which offers mathematical investigations and games for students of all ages, along with suggestions to help you coordinate those activities with your high school math curriculum.[‡]

♦ Explore a variety of challenge problems with Alcumus, the Art of Problem Solving's online learning system. Working out problems and reading other people's solutions (or watching the video) is a great way to learn math.[§]

♦ Delve into quadratics, trigonometry, and many other high school topics with James Tanton's delightful Global Math Project and G'Day Math video courses.[¶]

♦ Find new games to play (like Eleusis Express) on John Golden's Math Hombre Games webpage. Or create some mathematical beauty with the ideas on his Math × Art page.[**]

♦ And check out the other links in the "Favorite Resources" appendix in this book and the "Internet Math Resources" page on my Let's Play Math blog.[††]

Way back in the first chapter, we tried to imagine how a perfect teacher might help children learn math. But there is no perfect teacher. There is only you and me, and we have no superpowers. We can't save the world or solve the latest crisis of educational policy.

But we can teach our children to exercise their common sense as they interpret the stories of mathematics.

[†] *mathpickle.com*
mathpickle.com/unsolved-k-12
[‡] *nrich.maths.org*
nrich.maths.org/curriculum-secondary
[§] *artofproblemsolving.com/alcumus*
[¶] *globalmathproject.org*
gdaymath.com
[**] *mathhombre.blogspot.com/p/games.html*
mathhombre.blogspot.com/p/mathart.html
[††] *denisegaskins.com/internet-math-resources*

We can encourage them to practice algebraic thinking and develop problem-solving skills.

And through games like the ones in this book, we can prevent (or treat) math anxiety and build a positive attitude toward learning.

So what are we waiting for?

Let's play some math!

SECTION IV

Resources and References

Appendix A

Game-Playing Basics, From Set-Up to Endgame

When I was a child, I assumed that whatever I knew was common knowledge and anything I believed was common sense. Now I have grown up enough to realize how very much I do not know. So I understand how confusing new ideas (or new games) can be. Below I summarize everything I can think of that might be assumed-but-never-explained about playing games in the *Math You Can Play* series.

If you have a question I didn't answer, please send me an email: LetsPlayMath@gmail.com.

Math Concepts

Most of the *Math You Can Play* games build your students' skill at working with numbers in their heads. Some games focus on one or two concepts, while others cover a wide range of ideas. The latter are not necessarily better than the former.

Players

Most of the games are designed for two or more competing players, but a few also work as solitaire games. If a game relies on chance, it may not matter how many people are playing. But the more strategy involved in a game, the more likely it will work best as a two-player battle of minds.

When playing with a larger group, it may work better to split up and play separate games so players don't have too much idle time between their turns. To wait patiently can be difficult for an adult, so we shouldn't be surprised that it's hard for children.

If you are playing with a wide range of skill levels, avoid games that rely on speed or change them to allow each player adequate time to think.

Sometimes, you may allow extra turns to the younger students. For instance, in Concentration (Memory), you might let younger children turn up three cards instead of the usual two, giving them a better chance to find a pair that match.

Who Goes First?

Sometimes there is an advantage to going first in a game, so I often let the youngest player go first. You can randomize the turns by letting each player throw a die or draw one card from the deck, and then whoever gets the highest number goes first. In multi-player card games, whoever gets the lowest number is the dealer, and the player sitting to the dealer's left goes first.

Or follow the "pie rule"—one person slices the pie, and the other chooses which piece to take. For two-player strategy games: One player makes the first move, and then the other chooses whether to play second or to trade places.

Shuffle and Cut

In card or domino games, the players must mix the cards or tiles to ensure randomness. Any player may shuffle, but in card games, the dealer has the right to shuffle last. Players should not try to sneak a peek during the shuffle, and to take advantage of an accidentally revealed card is cheating.

The *riffle shuffle* (in which a deck of cards is split in half and then the halves are interlaced) takes plenty of practice, but it is the fastest way to randomize the cards. It takes about seven riffles to shuffle a deck fully. If you've never seen a riffle shuffle, search YouTube for a video that shows the technique. For example: "The Riffle Shuffle (in the hands)" by Pokerology.[†]

For young children, the easiest way to shuffle is domino style. Spread all the cards face down on the table, mix them around, and then stack them up again without looking.

If only the dealer shuffles the cards, it is polite to offer another player (usually whoever is to the dealer's right, opposite the direction of the deal) the chance to cut the deck. The player splits the deck into two parts, with at least four cards in each part, places the top part on the table without looking at it, and then stacks the other part on top of it. Thus neither player nor dealer can know the exact position of any card.

---

† *youtu.be/3oabnbtJRNQ*

The riffle shuffle is an efficient way to randomize a deck of cards.

Deal and Rotation of Play

In many card games, one person—the dealer—will hand out cards to each player in turn, going around the table in the direction of play. The dealer may give one card at a time, or two or more at once, but should deal the same to every player. Out of politeness and to avoid putting other players at a disadvantage, everyone should wait until all cards are dealt before picking up and looking at their hands.

My family plays by the tradition common in the United States that the deal and the players' turns go to the left (clockwise) around the table. I have read that many countries do the opposite, rotating play to the right. For the games in this book, direction does not matter, so use whichever seems comfortable to you.

Hand vs. Round vs. Game

We call the cards a player holds his or her hand. These are normally kept hidden from the other players until used in the game. Children often need a reminder to hold their hands close to their bodies so that the other players do not see their cards.

One complete section of a game, where every player has a turn or chance to play, may be called a hand or a round. Sometimes the terms are interchangeable, but for more complicated games it may take several hands to make a round and several rounds to finish a complete game.

Draw Pile (Stock) and Discard Pile

In games where players will need to refresh their hands, the rest of the deck (after cards have been dealt) is turned face down and placed on the table where everyone can reach. This is the *stock* or *draw pile*.

The unwanted cards from the players' hands are often turned face up next to the draw pile, either as a single stack or fanned out so all are visible. In some games, these discards may be available for other players to use in subsequent turns.

In many games for young children, we use a *fishing pond* in place of a draw pile. Turn all the cards face down and spread them out to form a roughly circular area where all players can reach. On their turns, players may choose any card. Discards should be mixed back into the pond before the next player's turn, so that nobody can remember their location.

Misdeal and Other Irregularities

When playing with children, you can almost guarantee that misdeals or exposed cards will happen. Some traditional games specify harsh penalties for such irregularities—think of old Western movies, where a game of poker could break into a bar fight or shootout over a simple mistake. In a family game, we can be more lenient. Our children must learn that cheating ruins the game for everyone, but there is no shame in an unintended error.

The dealer may give the wrong number of cards to a player or expose a card that other players are not supposed to see. Just fix the misdeal in a way that seems fair all around—for instance, by mixing the offending cards back into the deck or by reshuffling and starting over. If the players have looked at their hands before realizing they have too many cards, no one should choose which of his or her own cards to give back. Players can fan out their hands and let the dealer or another player who can't see the cards pick and discard the extras.

In the same way, anything wrong that happens in a game should be resolved in such a way as seems fair to every player. For example, if someone plays a card out of turn or makes an illegal play, the exposed card should stay face up on the table for use at the next legal opportunity. Or if the deck is

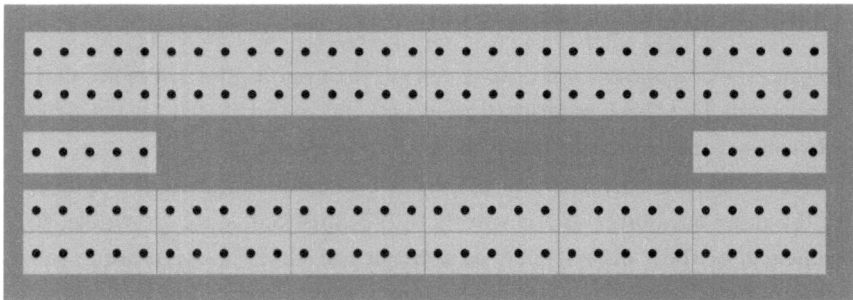

A 60-point cribbage board for two players.
Move the pegs away from yourself up your first row of thirty holes and then back down the second row. Use the holes in the middle for longer games, to count your trips around the board.

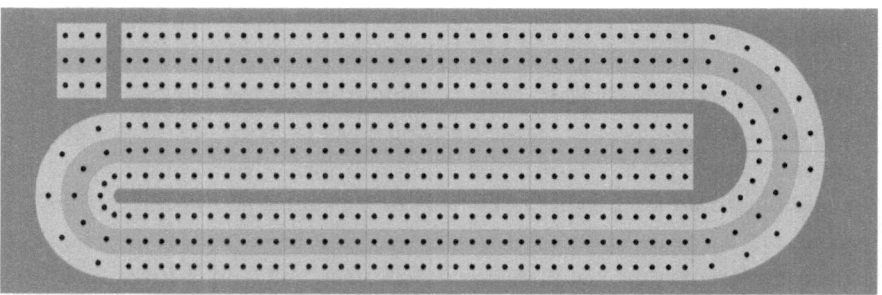

A three-player cribbage board for a game of 120 points. Follow your own line of holes around the loop to the end. If you use the three extra holes to count trips around the board, you can play longer games.

bad (perhaps the players discover that a few cards are missing), the current hand should start over with a new deck, but any points from previous rounds should stand.

Keeping Score

At our house, we often play for the next deal, rather than for points. My children enjoy having control over the game, so getting to deal is a treat for them. When we do play for points, the kids love to use poker chips to keep score: white = 1 point, red = 5 points, and blue = 10 points.

You could let your children practice money skills by using coins to keep score. Give one penny per point, with players trading in for higher coins as they progress, and the first player to collect $1 (or $5 or $10) wins the game.

Or you may use face cards and jokers as tallies in games where the winner of each hand gets a point. Give one tally card per point until they are gone, and whoever collects the most cards is the champion. Since there are three face cards in each of the four suits, this will make twelve hands, or fourteen with both jokers.

Or Try a Cribbage Board

You can often pick up cards, dice, dominoes, and poker chips at garage sales for next to nothing. A rarer discovery worth grabbing if you see one is a *cribbage board* (sometimes called a *crib board*). You don't have to know how to play cribbage to find this useful. It can be a great way to record points in many games.

Keep score using two small pegs in leapfrog fashion. Each hole represents a point, and the holes are arranged in groups of five for easy counting. To record your first score, count that many holes and place your first peg. For the second score, leave the first peg where it is and count beyond it, placing the second peg at your new total. For each succeeding set of points, leave the farthest-advanced peg in place to guard against losing count, and jump the other peg past it to the new total. The first player to *peg out*—that is, to reach the last hole—wins the game.

If you buy it used, your cribbage board may have lost its pegs. You can snip the sharp ends off round toothpicks or use wooden matchsticks as makeshift cribbage pegs.

Appendix B

A Few of My Favorite Resources

If you know of a fantastic math games resource I missed, please send me an email (LetsPlayMath@gmail.com). I appreciate your help!

Best-Loved Books

Most of these books should be available through your local library or via inter-library loan. Check for recreational games in the 793–795 range in the Dewey decimal system, and look for elementary education games at 372.

Avoid Hard Work! by Maria Droujkova, James Tanton, and Yelena McManaman

Camp Logic by Mark Saul and Sian Zelbo

Family Math by Jean Kerr Stenmark, Virginia Thompson, and Ruth Cossey

Games with Pencil and Paper by Eric Solomon

Hexaflexagons and Other Mathematical Diversions and other books by Martin Gardner

Math for Smarty Pants and *The 'I Hate Mathematics!' Book* by Marilyn Burns

Math Games and Activities from Around the World and other books by Claudia Zaslavsky

Mathematical Activities: A Resource Book for Teachers and *The Amazing Mathematical Amusement Arcade* and other books by Brian Bolt

Playing with Math: Stories from Math Circles, Homeschoolers, and Passionate Teachers edited by Sue VanHattum

This Is NOT a Maths Book and *This Is NOT Another Maths Book* by Anna Weltman

Online Games and Resources

The Internet overflows with a wide-ranging assortment of math websites. The list at my blog is much longer than this, and the "good intentions" folder of links I hope to add someday is longer still.

All the website links were checked before publication, but the Internet is volatile. If the website disappears, you can run a browser search for the author's name or article title. Or try entering the web address at the Internet Archive Wayback Machine.
archive.org/web/web.php

ALCUMUS: Art of Problem Solving's innovative online learning system adjusts to student performance to deliver appropriate problems and lessons.
artofproblemsolving.com/alcumus

BETTER EXPLAINED: Intuitive, often visual, explanations of high school math topics.
betterexplained.com

CUT THE KNOT INTERACTIVE: "Mathematics Miscellany and Puzzles," one of my all-time favorite sites. See also "Math Games and Puzzles, A Short Illustrated List."
cut-the-knot.org
cut-the-knot.org/games.shtml

DAILY TREASURE: Solve the logic puzzle to find the hidden gold.
4chests.blogspot.com

DESMOS GRAPHING CALCULATOR: Explore the relationships between equations and shapes, and try your hands at some of the Daily Desmos blog challenges.
desmos.com
dailydesmos.com

DON COHEN'S MAP OF CALCULUS FOR YOUNG PEOPLE: Hands-on activities featuring advanced ideas, for students of any age.
mathman.biz/html/map.html

DISCOVERING TRIGONOMETRY: A basic introduction to trigonometry, starting with sticks and shadows.
catcode.com/trig/index.html

ESTIMATION 180: "Building number sense one day at a time." How close can you guess? Why did you pick that number?
estimation180.com

Games and Math and Pam Sorooshian on Dice: Math itself is a game we play.
sandradodd.com/math/pamgames
sandradodd.com/math/pamdice

GeoGebra: Download software for playing with geometry and algebra, and the website offers a wealth of user-created instructional materials.
geogebra.org

How to Think like a School Math Genius: James Tanton's series of videos about five key principles for mathematical thinking for students approaching high school math.
jamestanton.com/?p=1097

Incompetech: Free online graph paper PDFs galore for any math game.
incompetech.com/graphpaper

Math Pickle: Videos introduce fun and challenging printable games and puzzles for K–12 students. Can your students solve the $1,000,000 problems?
mathpickle.com

Math Playground: A fun site for a variety of math games.
mathplayground.com

A New Algebra: Henri Picciotto offers a selection of interesting activities for algebra students.
mathedpage.org/new-algebra/new-algebra.html

Natural Math: Plenty of ideas for sharing rich math experiences with your children.
naturalmath.com

Nrich.maths.org: A wonderful source of math games and activities for all ages, with a theme that changes each month.
nrich.maths.org/public/index.php

Problem Solving Island: A variety of puzzles, from the book *Thinking Mathematically* and other sources, plus problem solving tips and sample student journal entries.
math.grin.edu/~rebelsky/ProblemSolving/index.html

Scratch: A programming language that makes it easy for students to create interactive stories, animations, games, music, and art.
scratch.mit.edu

Set Daily Puzzle: Logic puzzle game for all ages.
setgame.com/set/puzzle_frame.htm

TIM'S INTERACTIVE PUZZLE SOLUTION CENTER: A fun collection of "famous and other curious brain teasers" to solve online, some relatively easy and some quite challenging.
sakharov.net/puzzle

ULTIMATE LIST OF PRINTABLE MATH MANIPULATIVES AND GAMES: A treasure list from one of my favorite homeschooling blogs.
jimmiescollage.com/2011/04/ultimate-list-of-printable-math-manipulatives-games

VISUAL PATTERNS BLOG: Pick any design you like and practice recognizing, describing, and predicting the pattern.
visualpatterns.org

WHICH ONE? PUZZLERS: Based on Christopher Danielson's book *Which One Doesn't Belong?* Thought-provoking for math teachers and students alike. After you try these, make up your own puzzles to submit.
wodb.ca/index.html

WOULD YOU RATHER? MATH: "Asking students to choose their own path and justify it."
wouldyourathermath.com

Board and Card Games for Family Play

> *"Board games are a celebration of problem solving, and problem solving is at the heart of a quality mathematics education. The mathematics might be hidden, but I guarantee you that it will be there."*
>
> —GORDON HAMILTON

In addition to the classics of strategy—backgammon, chess, mancala, Othello/Reversi, Pente, and so on—your family can enjoy (and learn from) many modern games. For example:

7 WONDERS: Lead an ancient civilization as it rises from its barbaric roots to become a world power.

BLOKUS: Strategy game for up to four players.

CARCASSONNE: Lay down your tiles to create a landscape based on southern France.

CITADELS: Bluffing, deduction, and city-building set in a medieval world.

DVONN: An abstract strategy game based on moving pieces to make the largest stack.

For Sale: Buy and sell real estate to amass your fortune.

Forbidden Island: Capture four sacred treasures from the ruins of this perilous paradise.

King of Tokyo: Mutant monsters, gigantic robots, and other aliens vie for the right to rule the city.

Labyrinth: Find a path to collect your treasures, but watch out—the maze shifts and changes on every turn.

Lost Cities: Explore the world in search of ancient civilizations.

Love Letter: A short, simple game that combines luck and strategy.

Memoir '44: Test your strategic skills as you refight the battles of World War II.

Mr. Jack: Jack the Ripper is loose in Whitechapel, and it's up to you to stop him.

Munchkin Quest: Explore the dungeon and battle monsters for power and treasure.

Power Grid: Acquire raw materials, upgrade your power plants, and expand your network to more cities.

Quarto: Four-in-a-row strategy game.

Quirkle: Strategically match colors and shapes to build up your score.

Quoridor: Move your pawn through the maze, and block the other players.

Set: Visual perception card game.

Settlers of Catan: A trading and building game set in a mythical world.

Smash Up: Easy to learn, fun to play, and always different.

Splendor: As a Renaissance merchant, you must acquire mines and transportation, hire artisans, and woo the nobility.

Stone Age: Gather resources to feed and shelter your tribe.

Ticket to Ride: A cross-country train adventure. How many cities can you visit?

Zeus on the Loose: Use addition, subtraction, and strategic thinking to capture the runaway god.

Zooloretto: Plan carefully to attract as many visitors as possible to your zoo.

Appendix C

Quotes and Reference Links

I LOVE QUOTATIONS. EVERYTHING I could ever want to say has probably been said sometime by someone else (who did not think of it first, either). At least a few of those people had a wonderful way with words.

Some of the quotations in this book are from my own reading. Others are gleaned from two websites I visit often to browse: Furman University's Mathematical Quotation Server and the mathematics collection at Wikiquote.[†]

All the website links were checked in late 2020 (dates of original publication are included if the website provided them), but the Internet is volatile. If the website disappears, you can run a browser search for the author's name or article title. Or try entering the web address at the Internet Archive Wayback Machine.[‡]

ANONYMOUS. "We do not stop playing because we grow old..." Attributed to Benjamin Franklin, Oliver Wendell Holmes, and George Bernard Shaw, among others. Choose your favorite sage.

ANONYMOUS MATHCHIQUE. "I love playing math games..." from a comment on the Let's Play Math blog.
denisegaskins.com/2008/05/13/game-function-machine

ANONYMOUS MATH MOM. "Figurate Numbers, and the Unsummables," Ramblings of a Math Mom blog, September 28, 2007.
mathmomblog.wordpress.com/2007/09/28/figurate-numbers-and-the-unsummables

—. "Middle Schoolers and the Unsummables," Ramblings of a Math Mom blog, October 12, 2007.
mathmomblog.wordpress.com/2007/10/12/middle-schoolers-and-the-unsummables

ANONYMOUS POKEROLOGY. "The Riffle Shuffle (in the hands)," YouTube,

---

[†] *math.furman.edu/~mwoodard/mquot.html*
en.wikiquote.org/wiki/Mathematics
[‡] *archive.org/web/web.php*

December 15, 2010.
youtu.be/3oabnbtJRNQ

Antonick, Gary. "The Tax Collector," *The New York Times* Numberplay blog, April 13, 2015.
wordplay.blogs.nytimes.com/2015/04/13/finkel-4

Batterson, Jason. *Competition Math for Middle School*, self-published, 2009. After creating the AGmath website, Batterson joined the Art of Problem Solving team in 2010 to lead development of the Beast Academy series.
agmath.com

—. "Counting and Probability," AGmath website.
agmath.com/media//DIR_57727/1$20Counting.pdf

—. "From Counting to Probability," AGmath website.
agmath.com/media//DIR_57727/Probability.pdf

Bishop, Jessica Pierson and Randolph Philipp. "A CGI-Approach to Integers: Helping Teachers Structure Their Intuitive Knowledge About Children's Understandings of Negative Numbers," plenary address to the Teachers Development Group Leadership Seminar (transcript and Handout 3), Portland, Oregon, February 14, 2014.
sci.sdsu.edu/CRMSE/projectz/presentations.html

Booker, George. "Algebraic Thinking: Generalising Number and Geometry to Express Patterns and Properties," Mathematical Association of Victoria 46th Annual Conference, 2009.
research-repository.griffith.edu.au/handle/10072/30846

Boole, Mary Everest. "Algebra deals not with laws of number..." from *Lectures on the Logic of Arithmetic,* Clarendon Press, 1903. Available at the Internet Archive.
archive.org/details/lecturesonlogic02boolgoog

—. "Dealing logically with the fact of our own ignorance..." from *Philosophy and Fun of Algebra,* C. W. Daniel, 1909. Available at the Internet Archive.
archive.org/details/philosophyandfun030851mbp

Borenson, Henry. Hands-On Equations curriculum.
borenson.com

Brickhill, Paul. *The Great Escape,* W. W. Norton & Co, 1950.

Broug, Eric. "Classroom Resources," School of Islamic Geometric Design website. Broug is the author of several books on Islamic geometric design and

gives workshops and lectures around the world.
sigd.org/resources/classroom-resources-islamic-geometric-design

BURNS, MARILYN. *About Teaching Mathematics,* 3rd ed., Math Solutions Publications, 2007. Burns has been a teacher, author, and workshop speaker for more than fifty years and is the founder of Math Solutions.

—. "The Game of Pathways," Marilyn Burns Math Blog, March 10, 2016.
marilynburnsmathblog.com/wordpress/the-game-of-pathways

BUTLER, DAVID. "I was struck by how quickly…" from "Digit Disguises," Making Your Own Sense blog, September 21, 2019. Butler is a lecturer in the Maths Learning Centre at the University of Adelaide, Australia.
blogs.adelaide.edu.au/maths-learning/2019/09/21/digit-disguises

CANTOR, GEORG. "A set is a Many…" quoted by Rudy Rucker in *Infinity and the Mind,* Princeton Science Library, 1995.
rudyrucker.com/infinityandthemind

CARDONE, TINA. *Nix the Tricks: A Guide to Avoiding Shortcuts that Cut Out Math Concept Development,* self-published, 2013.
nixthetricks.com

CARMICHAEL, ALLISON, AND MARTIN WOODS. "Printable Algebra Game," Parent Concept website, Indigoextra, Ltd. Carmichael and Woods provide information to parents considering home education in France.
parentconcept.com/printable-algebra-game

CHILDREN'S TELEVISION WORKSHOP. "But Who's Multiplying?" *Square One TV,* first season, 1987. Episodes available on YouTube.
tv.com/shows/square-one-tv/episodes
youtube.com/watch?v=p7C2w2WAeVU
youtube.com/watch?v=V-sBHbQBTgM
youtube.com/watch?v=iB9ff_NYJoY

CLEVELAND, JAMES R. "Fighting for the Center," Roots of the Equation blog, July 27, 2015. Cleveland is a high school math teacher in New York City and a Math for America master teacher.
rootsoftheequation.wordpress.com/2015/07/27/fighting-for-the-center

CMGLEE. "Regular Star Polygons.svg," Wikimedia Commons, 2014.
commons.wikimedia.org/wiki/File:Regular_star_polygons.svg

COPPING, JASPER. "The Great Escape Failed but It Was Worth It, Say Veterans 70 Years On," *The Telegraph* website, March 12, 2014.
telegraph.co.uk/history/world-war-two/10693186/The-Great-Escape-failed-but-it-was-worth-it-say-veterans-70-years-on.html

DANIELSON, CHRISTOPHER. "I need to tell you a secret…" from "How I Learned to Love Middle School Geometry," MiddleWeb website, December 11, 2016. With his blogs and other projects, Danielson helps parents and teachers understand how children understand math.
middleweb.com/33478/how-i-learned-to-love-middle-school-geometry

—. *Talking Math with Your Kids*, self-published, 2013.
talkingmathwithkids.com

—. "Hierarchy of Hexagons," Overthinking My Teaching blog, 2011–2014.
christopherdanielson.wordpress.com/?s=hexagons
bit.ly/hexagonspdf

DEVLIN, KEITH. "Hacker and I have…" from "Andrew Hacker and the Case For and Against Algebra," Huffington Post website, February 29, 2016. Devlin is a mathematician, author, video-game designer, "The Math Guy" on National Public Radio, and director of the Stanford Mathematics Outreach Project.
huffpost.com/entry/andrew-hacker-and-the-cas_b_9339554

—. "How Do You Read '−3'?" Huffington Post website, March 12, 2012.
huffpost.com/entry/how-do-you-read-3_b_1338163

—. "The Math Myth that Permeates 'The Math Myth,'" Devlin's Angle blog, March 1, 2016.
devlinsangle.blogspot.com/2016/03/the-math-myth-that-permeates-math-myth.html

DOMORADZKI, JOHN. Son of Amber Domoradzki, author of *Misbehavior Is Growth: An Observant Parent's Guide to the Toddler Years*, self-published, 2018.

DONNE, JOHN. "No man is an island…" from "Meditation 17," *Devotions upon Emergent Occasions*, 1623. Excerpt available at Wikisource, full text at Project Gutenberg.
en.wikisource.org/wiki/Meditation_XVII
gutenberg.org/ebooks/23772

DUDENEY, HENRY ERNEST. *The Canterbury Puzzles*, E.P. Dutton and Co., 1908. Available at the Internet Archive.
archive.org/details/117770747

EDUCATION DEVELOPMENT CENTER TEAM. SolveMe Puzzles website.
solveme.edc.org

ERNEST, JAMES. "Gold Digger," Cheapass Games website. Printable rules and playing cards.
cheapass.com/free-games/gold-digger

EVES, HOWARD. "A good problem should be more…" from *An Introduction to the History of Mathematics*, Thomson Learning, Inc, 6th ed., 1990.

FINKEL, DAN. "Of all the myths…" from "Grids, Polygons and Acute Angles," *The Hindu in School*, October 8, 2017. Shared in "A Mathematician at Play: Puzzle 1," Math for Love blog. Finkel is the founder of Math for Love and the creator of mathematical puzzles, curriculum, and games, including the best-selling Prime Climb and Tiny Polka Dot.
mathforlove.com/wp-content/uploads/2019/10/HinduPuzzle1.pdf
mathforlove.com/2018/01/mathematician-at-play-puzzle-1

FULTON, BRAD AND BILL LOMBARD. "The tax collector usually wins…" from "The Tax Collector," *Simply Great Math Activities*, Teacher to Teacher Press. Printable file shared on Mr. L's Math blog (sadly defunct), rescued by the Internet Archive Wayback Machine.
tinyurl.com/tax-collector-pdf

—. "Four-in-a-Row," *Simply Great Math Activities*, Teacher to Teacher Press. Mr. L's Math blog archive.
tinyurl.com/four-in-a-row-pdf

—. "Games and Puzzles that Reach the Kids and Teach the Standards," notes from the California Math Council conference at Asilomar, December 2009.
tinyurl.com/games-and-puzzles-handout (PDF)

GALE, DAVID. "A Curious Nim-type Game," *The American Mathematical Monthly*, vol. 81, issue 8, 1974.

GALILEI, GALILEO. "The universe…" from *Il Saggiatore (The Assayer)*, 1623, as translated by Stillman Drake.
web.stanford.edu/~jsabol/certainty/readings/Galileo-Assayer.pdf

GARDNER, MARTIN. "Pentominoes and Polyominoes: Five Games and a Sampling of Problems," Mathematical Games column in *Scientific American* magazine, October 1965.

—. "Sim, Chomp and Race Track: New Games for the Intellect (and Not for Lady Luck)," Mathematical Games column in *Scientific American* magazine, January 1973.

GASKINS, DENISE. *Let's Play Math: How Families Can Learn Math Together, and Enjoy It*, Tabletop Academy Press, 2016.

—. *Prealgebra & Geometry Printables*, Tabletop Academy Press website.
tabletopacademy.net/free-printables

—. "Backwards Math," Let's Play Math blog, April 3, 2008.
denisegaskins.com/backwards-math

—. "Internet Math Resources," Let's Play Math website.
denisegaskins.com/internet-math-resources

—. "Math Game: War with Special Decks," Let's Play Math blog, May 6, 2020.
denisegaskins.com/2020/05/06/math-game-war-with-special-decks

—. "More Backwards Math," Let's Play Math blog, September 3, 2008.
denisegaskins.com/more-backwards-math

—. "My Favorite Math Games," Let's Play Math blog, January 7, 2017.
denisegaskins.com/my-favorite-math-games

—. "Year Game," Let's Play Math website.
denisegaskins.com/tag/year-game

GAUSS, CARL FRIEDRICH. "Mathematics is the queen…" quoted by Wolfgang Sartorius von Waltershausen in *Gauss zum Gedächtniss*, 1856.

GOLDEN, JOHN. "Be careful! There are a lot of useless games…" from the "Math Games for Skills and Concepts" handout on the Games Reference Page, Math Hombre blog. Golden helps train future math teachers as an associate professor at Grand Valley State University, and trains the rest of us through the posts on his blog. I love his Math × Art webpage!
faculty.gvsu.edu/goldenj/GameshandoutHS.pdf
mathhombre.blogspot.com/p/games.html
mathhombre.blogspot.com/p/mathart.html

—. "Area Battle," Math Hombre blog, October 7, 2011.
mathhombre.blogspot.com/2011/10/area-battle.html

—. "Area Block," Math Hombre blog, October 25, 2009.
mathhombre.blogspot.com/2009/10/area-block.html

—. "Four Corners," Math Hombre blog, May 2, 2014.
mathhombre.blogspot.com/2014/05/four-corners.html

—. "Greater Than," Math Hombre blog, September 30, 2012.
mathhombre.blogspot.com/2012/09/greater-than.html

—. "Guess My Rule," Math Hombre blog, May 10, 2012.
mathhombre.blogspot.com/2012/05/guess-my-rule.html

—. "Integer Games," Math Hombre blog, February 4, 2011.
mathhombre.blogspot.com/2011/02/integer-games.html

—. "Linear War," Math Hombre blog, July 4, 2011.
mathhombre.blogspot.com/2011/07/linear-war.html

—. "Making Whoopee at TMC15," Math Hombre blog, August 19, 2015.
mathhombre.blogspot.com/2015/08/making-whoopee-at-tmc15.html

—. "Multiplying Fractions, Times Three," Math Hombre blog, May 13, 2010.
mathhombre.blogspot.com/2010/05/multiplying-fractions-times-three.html

—. "Pentiremeter," Math Hombre blog, January 27, 2017.
mathhombre.blogspot.com/2017/01/pentiremeter.html

—. "Playing Math," Math Hombre blog, July 3, 2010.
mathhombre.blogspot.com/2010/07/playing-math.html

—. "Power Up," Math Hombre blog, May 16, 2011.
mathhombre.blogspot.com/2011/05/power-up.html

—. "Product Game … Again!" Math Hombre blog, March 25, 2011.
mathhombre.blogspot.com/2011/03/product-game-again.html

—. "Same Game, Different Grade," Math Hombre blog, July 24, 2017.
mathhombre.blogspot.com/2017/07/same-game-different-grade.html

GOLOMB, SOLOMON W. *Polyominoes: Puzzles, Patterns, Problems, and Packings*, Princeton University Press; 2nd ed., 1994.

GREENE, JOSHUA. "My Closest Neighbor," Three J's Learning blog, 2017. Greene is a math education hobbyist and president of the PROMYS Foundation.
3jlearneng.blogspot.com/2017/01/my-closest-neighbor-fraction-game.html
3jlearneng.blogspot.com/2017/01/closest-neighbor-one-on-one.html
3jlearneng.blogspot.com/2017/02/perfect-play-for-my-closest-neighbor.html

HACKER, ANDREW. *The Math Myth: And Other STEM Delusions,* The New Press, 2016. Hacker is an emeritus professor of political science at Queens College and the author of many books.

—. "Is Algebra Necessary?" *The New York Times,* July 29, 2012.
nytimes.com/2012/07/29/opinion/sunday/is-algebra-necessary.html

HAINES, KENT. "I love this game because…" from "Integer Solitaire," The Process Column blog, February 19, 2016. Kent Haines is a math teacher and author of the *Games for Young Minds* online newsletter.
web.archive.org/web/20190313055300/http://www.kenthaines.com/blog/2016/2/19/integer-solitaire
gamesforyoungminds.com

HAMILTON, GORDON. "Board games are a celebration…" from "Commercial

Games," YouTube video, December 25, 2011. Hamilton is the founder of the Math Pickle website and creator of the board game Santorini.
youtu.be/J8geFOkOUbU

—. "Unsolved K-12," Math Pickle website, 2013.
mathpickle.com/unsolved-k-12

—. "Venn Diagram Puzzles," Math Pickle website, 2011.
mathpickle.com/project/venn-diagram-puzzles

HAMKINS, JOEL DAVID. "Doubled, Squared, Cubed: A Math Game for Kids or Anyone," Joel David Hamkins website, October 21, 2013. Hamkins is a professor of logic and the Sir Peter Strawson Fellow at University College, Oxford.
jdh.hamkins.org/doubled-squared-cubed-math-game-for-kids

—. "The Rule-Making Game," Joel David Hamkins website, October 21, 2013.
jdh.hamkins.org/the-rule-making-game

HUFF, DARRELL. *How to Lie with Statistics,* Penguin Books, 1973. Available at the Internet Archive.
archive.org/details/HowToLieWithStatistics

KAMII, CONSTANCE, WITH LESLIE BAKER HOUSMAN. *Young Children Reinvent Arithmetic: Implications of Piaget's Theory,* 2nd ed., Teachers College Press, 2000.

— WITH LINDA LESLIE JOSEPH. *Young Children Continue to Reinvent Arithmetic, 2nd Grade: Implications of Piaget's Theory,* 2nd ed., Teachers College Press, 2004.

— WITH SALLY JONES LIVINGSTON. *Young Children Continue to Reinvent Arithmetic, 3rd Grade: Implications of Piaget's Theory,* Teachers College Press, 1994.

KAWAS, TERRY. "Factor Blaster," MathWire blog, September 26, 2010. Kawas founded the MathWire site to provide classroom teachers with activities and worksheets that support a constructivist approach to learning mathematics.
mathwire.blogspot.com/2010/09/factor-blaster.html

—. "Shape Capture Game," MathWire website, January 2006.
mathwire.com/games/shapecapture.pdf

KAYE, PEGGY. "Children learn more math and enjoy math more…" from *Games for Math,* Pantheon Books, 1988. If you're homeschooling young children, be sure to check out the other books in Kaye's *Games for…* series.
peggykaye.com

Kilpatrick, William Heard. "[Algebra is] an intellectual luxury," quoted by Samuel Tenenbaum in *William Heard Kilpatrick: Trail Blazer In Education*, Harper & Brothers Publishers, 1951.
archive.org/details/in.ernet.dli.2015.264846

Laib, Jenna. "One of My Favorite Games: Number Boxes," Embrace the Challenge blog, May 29, 2019. Laib works as a K–8 math specialist near Boston.
jennalaib.wordpress.com/2019/05/29/one-of-my-favorite-games-number-boxes

—. Slow Reveal Graphs website.
slowrevealgraphs.com

Leo, Lucinda. "With any curriculum there is the temptation…" from "Things I've Learned About Homeschooling," Navigating by Joy blog, December 10, 2013. Leo is an English mom who blogs about her family's unschooling adventures.
navigatingbyjoy.com/2013/12/10/3-things-ive-learned-homeschooling-2013

Lombard, Bill, and Brad Fulton. "The tax collector usually wins…" from "The Tax Collector," *Simply Great Math Activities,* Teacher to Teacher Press. Printable file shared on Mr. L's Math blog (sadly defunct), rescued by the Internet Archive Wayback Machine.
tinyurl.com/tax-collector-pdf

—. "Four-in-a-Row," *Simply Great Math Activities,* Teacher to Teacher Press. Printable file originally from Mr. L's Math blog archive.
tinyurl.com/four-in-a-row-pdf

—. "Games and Puzzles that Reach the Kids and Teach the Standards," notes from the California Math Council conference at Asilomar, December 2009.
tinyurl.com/games-and-puzzles-handout (PDF)

Lyon, Jack. "It did a lot for morale…" quoted by Jasper Copping in "The Great Escape Failed but It Was Worth It, Say Veterans 70 Years On," The Telegraph website, March 12, 2014.
telegraph.co.uk/history/world-war-two/10693186/The-Great-Escape-failed-but-it-was-worth-it-say-veterans-70-years-on.html

McLeod, John. "Card Game Rules: Card Games and Tile Games from Around the World," Pagat website. Pagat is a collection of card game rules and history tidbits.
pagat.com

Meyer, Dan. "Asilomar #5: Michael Serra," dy/dan blog, December 7, 2008.

Meyer taught high school math (to students who didn't like high school math) and now serves as the Chief Academic Officer at Desmos.
blog.mrmeyer.com/2008/asilomar-5-michael-serra

—. "Tiny Math Games," dy/dan blog, April 16, 2013.
blog.mrmeyer.com/2013/tiny-math-games

MOSES, ROBERT P. (Bob). "Education is still basically Jim Crow…" quoted by Christopher Connelly in "To '60s Civil Rights Hero, Math Is Kids' Formula for Success," NPR Morning Edition, August 1, 2013. Moses founded The Algebra Project to promote math education for students in underserved schools.
npr.org/sections/codeswitch/2013/08/02/206813091/
  to-60s-civil-rights-hero-math-is-kids-formula-for-success
algebra.org

— AND CHARLES E. Cobb, Jr. *Radical Equations — Civil Rights from Mississippi to the Algebra Project,* Beacon Press, 2001.

— AND ED DUBINSKY. "Mathematics joins reading and writing…" in "Philosophy, Math Research, Math Ed Research, K–16 Education, and the Civil Rights Movement: A Synthesis," *Notices of the American Mathematical Society,* March 2011.
ams.org/notices/201103/rtx110300401p.pdf

NCTM ILLUMINATIONS TEAM. "Product Game," Illuminations website.
illuminations.nctm.org/Activity.aspx?id=4213

NGUYEN, FAWN. "Hotel Snap," Finding Ways blog, December 11, 2013. Nguyen is a Teacher on Special Assignment with Rio School District in Oxnard, California, and blogs about her lessons and classroom teaching.
fawnnguyen.com/hotel-snap

—. Visual Patterns website.
visualpatterns.org

NRICH TEAM. "Coordinate Cunning," Nrich Enriching Mathematics website. The Nrich site is a delightful collection of playful math activities and teaching tips.
nrich.maths.org/1279

—. "Factors and Multiples Game," Nrich Enriching Mathematics website.
nrich.maths.org/5468

—. "Secondary Curriculum," Nrich Enriching Mathematics website.
nrich.maths.org/curriculum-secondary

—. "Stop or Dare," Nrich Enriching Mathematics website.
nrich.maths.org/1193

—. "Strike It Out," Nrich Enriching Mathematics website.
nrich.maths.org/6589

NUNES, TEREZINHA, AND PETER BRYANT. *Children Doing Mathematics,* Wiley-Blackwell, 1996.

PHILIPP, RANDOLPH, AND JESSICA PIERSON BISHOP. "A CGI-Approach to Integers: Helping Teachers Structure Their Intuitive Knowledge About Children's Understandings of Negative Numbers," plenary address to the Teachers Development Group Leadership Seminar (transcript and Handout 3), Portland, Oregon, February 14, 2014.
sci.sdsu.edu/CRMSE/projectz/presentations.html

PIERCE, ROD. "Curriculum," Math Is Fun website. Pierce's site aims to explain K–12 mathematics in an enjoyable and easy-to-learn manner for both students and their parents.
mathsisfun.com/links/index-curriculum.html

—. "Vectors," Math Is Fun website.
mathsisfun.com/algebra/vectors.html

PLATO. "There should be no element of slavery in learning…" from *The Republic.* Unattributed translation quoted by Julian Fleron at the (now mostly dead) Mathematical and Educational Quotation Server at Westfield State University.
web.archive.org/web/20150929090318/http://www.westfield.ma.edu/math/faculty/fleron/quotes/viewquote.asp?letter=p

POLYA, GEORGE. "The teacher who wishes to serve…" from *Mathematical Discovery: On Understanding, Learning, and Teaching Problem Solving,* John Wiley & Sons, Combined Edition, 1981. Available at the Internet Archive.
archive.org/details/GeorgePolyaMathematicalDiscovery

POSITIVE ENGAGEMENT PROJECT. *Acing Math (One Deck at a Time): A Collection of Math Games,* The Positive Engagement Project website.
pepnonprofit.org/uploads/2/7/7/2/2772238/acing_math.pdf

POST, SONYA. "Math develops reasoning skills…" Facebook post, January 18, 2019. Former "math witch" and author of *Hands-On Learning with Gattegno,* Post equips homeschooling parents to understand and teach math.
facebook.com/nomathfears/posts/2189581941259597

—. "Exhausting Relationships: Learning the Rules of Algebra from the Beginning," Arithmophobia No More blog, 2017.
arithmophobianomore.com/exhausting-relationships

—. "Substitution Game — Forget the Worksheets," Arithmophobia No More blog, 2016.
arithmophobianomore.com/substitution-game-forget-worksheets

REULBACH, JULIE. "Math Games Collection on Google Docs — Add Your Game Today!" I Speak Math blog, November 18, 2013. Reulbach teaches high school math at a private school in North Carolina and is a speaker and Desmos Fellow.
ispeakmath.org/2013/11/18/
   math-games-collection-on-google-docs-add-your-game-today

SALOMON, PAUL. "Imbalance Problems," Lost in Recursion blog, 2013. Salomon is a mathematical artist, teacher, and founder of Art of Math STL.
lostinrecursion.wordpress.com/resources/imbalance-problems

SAWYER, W. W. "Earlier we considered…" from *Mathematician's Delight*, Penguin Books, Ltd., 1943. Any book by Sawyer is well worth reading. Available at the Internet Archive.
archive.org/details/MathematiciansDelight

SCHUSTER, LAINIE. *Enriching Your Math Curriculum, Grade 5: A Month-to-Month Guide*, Math Solutions, 2010.
mathsolutions.com/documents/978-1-935099-02-4_NL36_L1.pdf

SHORE, CHRIS. Clothesline Math website. Shore is a high school math teacher, coordinator of curriculum and instruction, speaker, and author of The Math Projects Journal website.
clotheslinemath.com
mathprojects.com

SINGH, MARINA. "The Great Escape," MathCurious blog, April 2, 2020. Marina Singh is a primary school teacher from Vancouver, Canada. With the help of her three boys, she creates and shares math puzzles, games, books, and teaching ideas.
mathcurious.com/2020/04/02/the-great-escape

SOROOSHIAN, PAM. "Mathematicians don't sit around…" from the old Unschooling Discussion Yahoo group, quoted by Sandra Dodd in "Games and Math," Unschoolers and Mathematics website. Sorooshian is a mother, college economics professor, and unschooling advocate.
sandradodd.com/math/pamgames

—. "Pam Sorooshian on Dice," Sandra Dodd's Unschoolers and Mathematics website.
sandradodd.com/math/pamdice

STEWARD, DON. "Three Operations, Expressions," Median blog, January 28, 2020. Steward creates math teaching resources and creative math tasks for middle and high school topics.
donsteward.blogspot.com/2020/01/three-operations-expressions.html

SULLIVAN, CHRISTINE. "My Closest Neighbor (Estimating with Fractions)," A Sea of Math blog, August 4, 2014. Sullivan is a former middle school math teacher.
a-sea-of-math.blogspot.com/2014/08/my-closest-neighbor-estimating-with.html

TANTON, JAMES. "In grade five I asked…" from "Why Is Negative Times Negative Positive?" Thinking Mathematics website, September 22, 2009. Tanton is an author, former teacher, and a founder of the Global Math Project.
jamestanton.com/?p=353
globalmathproject.org

—. "Why is negative times negative positive? Part I," YouTube, September 22, 2009.
youtu.be/nJqyr8h22nY

—. "Why is negative times negative positive? Part II," YouTube, September 22, 2009.
youtu.be/eV6iYvd4KS0

—. "Permutations and Combinations," G'Day Math website, 2014.
gdaymath.com/courses/permutations-and-combinations

VANHATTUM, SUE. "We say math is a language…" from "When Math Tells a Story," Math Mama Writes blog, February 12, 2019. VanHattum is a math professor at Contra Costa College, math circle leader, and blogger.
mathmamawrites.blogspot.com/2019/02/when-math-tells-story.html

—. "You can think of puzzles…" from "Parents and Kids Together: Learning in Community" in *Playing with Math: Stories from Math Circles, Homeschoolers, and Passionate Teachers,* edited by Sue VanHattum, Natural Math, 2015.
playingwithmath.org

WAY, JENNI. "Games can allow children to operate…" from "Learning Mathematics Through Games Series: 1. Why Games?" Nrich Enriching Mathematics website. Way is a math education researcher and associate professor at the University of Sydney.
nrich.maths.org/2489

WEDD, NICK, AND JOHN MCLEOD. "Mechanics of Card Games," Pagat website, May 15, 2009.
pagat.com/mech.html

WIKIPEDIA CONTRIBUTORS. "Central tendency," Wikipedia Internet Encyclopedia.
en.wikipedia.org/wiki/Central_tendency

—. "Pig (dice game)," Wikipedia Internet Encyclopedia.
en.wikipedia.org/wiki/Pig_%28dice_game%29

—. "Four Fours," Wikipedia Internet Encyclopedia.
en.wikipedia.org/wiki/Four_fours

—. "Gomoku," Wikipedia Internet Encyclopedia.
en.wikipedia.org/wiki/Gomoku

—. "The Great Escape (book)," Wikipedia Internet Encyclopedia.
en.wikipedia.org/wiki/The_Great_Escape_(book)

—. "Krypto (game)," Wikipedia Internet Encyclopedia.
en.wikipedia.org/wiki/Krypto_(game)

—. "Pentomino," Wikipedia Internet Encyclopedia.
en.wikipedia.org/wiki/Pentomino

—. "Pie Rule," Wikipedia Internet Encyclopedia.
en.wikipedia.org/wiki/Pie_rule

—. "Pig (dice game)," Wikipedia Internet Encyclopedia.
en.wikipedia.org/wiki/Pig_%28dice_game%29

—. "Racetrack (game)," Wikipedia Internet Encyclopedia.
en.wikipedia.org/wiki/Racetrack_(game)

WOODS, MARTIN, AND ALLISON CARMICHAEL. "Printable Algebra Game," Parent Concept website, Indigoextra, Ltd. Woods and Carmichael provide information to parents considering home education in France.
parentconcept.com/printable-algebra-game

ZASLAVSKY, CLAUDIA. "Language should be part of the activity..." from *Preparing Young Children for Mathematics: A Book of Games with Updated Book, Game and Resource Lists*, Schocken Books, 1986. Any book by Zaslavsky is well worth reading.

Index

A
absolute value, 54, 60
activities
    clothesline math, 50
    hexagon sort, 169
    pentominoes, 149
    polite numbers, 20
    star polygons, 142
    toy tug of war, 85
AGmath, 202, 218
Alcumus, 203, 213
algebra, 199
    as a puzzle, 56
    as a secret code, 45
    civil rights, 201
    constants, 80
    definitions of, 200
    equations, 85, 94, 112
    equivalence, 100
    expressions, 94, 100, 112
    factoring quadratic equations, 92
    functions, 108
    identities, 112
    imagination and, 83
    inequalities, 94, 112
    intellectual luxury, 199
    playing with ideas, 118
    substitution, 100, 112
    telling stories, 201
    variables, 80, 108
    with words, 90
Algebra Match, 113
Algebra Multi-Match, 114
Algebra-Tac-Toe, 115
Anonymous, 13, 217
Area Battle, 156
Area Block, 153
Area Dice, 155
area model, 70
arithmetic
    answer words, 47, 91
    averages, 68
    benchmark numbers, 37
    counting, 188
    exponential growth, 120, 124, 131
    factors and multiples, 25
    fractions, 37, 39, 54, 105
    inverse operations, 55, 57, 106
    multiplication with area model, 70
    natural numbers, 20
    negative numbers, 50
    number line, 51, 164, 170
    operations, 47
    order of operations, 100, 105
    prime vs composite, 26
    scale, 52
    squares and cubic numbers, 30
    statistics, 68
    symmetry, 195
    unary operator, 52
    Venn diagrams, 40
    zero, 51, 60, 124
art and math, 142, 146, 148, 203
Avengers, The, 95

B
Baldwin, Emma, 84
base (of exponent), 120
base (of prism), 160

Batterson, Jason, 187, 202, 218
benchmark numbers, 37
benefits of math games, 3, 5, 7
Bishop, Jessica, 49, 218
Boole, Mary Everest, 200, 218
Borenson, Henry, 85
Brickhill, Paul, 31, 218
Broug, Eric, 148, 218
Burns, Marilyn, 82, 212, 219
Butler, David, 48, 219

C

Canterbury Puzzles, The, 152, 220
Cantor, Georg, 40, 219
Captain Victory vs the Snake Lady, 123
card games
    Consecutive Capture, 53
    Exponent Pickle, 127
    Fight for the Center, 66
    Great Escape I, 29
    Great Escape II, 32
    Greater Than, 93
    Hit Me, 59
    Honeycomb, 73
    Integer Solitaire, 64
    Integer War, 76
    Krypto, 132
    Krypto Insanity, 136
    My Nearest Neighbor, 35
    Number Yoga, 134
    Operations, 103
    Power Krypto, 134
cards, types of, 13
Carmichael, Allison, 115, 219
Cartesian coordinates, 164
casework, 193, 195
Cavalieri, Bonaventura, 160
choose to believe, 58
civil rights, 201
classroom, using games in, 7
Cleveland, James, 69, 219
Clothesline Math, 50, 58, 228
combinatorics, 188
common sense, 49, 51, 199, 202
Competition Math for Middle School, 187, 192, 218

complexifying equations, 99, 118
composite number, 26
concave vs convex, 142
conjecture, 22, 145, 169
Consecutive Capture, 53
conversation, benefits of, 8, 55
conversational games
    Function Machine, 109
    Number Riddles, 43
    Number That Must Not Be Named, 138
    What Two Numbers?, 91
convex vs concave, 142
cooperative games
    Exhaust the Relationships, 116
    Factors and Multiples Cooperative, 24
    Great Escape II, 32
    Integer Solitaire, 64
    Number That Must Not Be Named, 138
    Substitution Game, 99
Coordinate Gomoku, 171
counting, 188
creating game variations, 9
cribbage board, 210
cubic numbers, 30
Cuisenaire rods, 116

D

Danielson, Christopher, 8, 141, 169, 215, 220
data, 68
Decimator, 39
degrees vs radians, 179
delta (change), 182
Descartes, René, 164
Despicable Me, 54
Devlin, Keith, 51, 199, 220
dice games
    Algebra Match, 113
    Algebra Multi-Match, 114
    Algebra-Tac-Toe, 115
    Area Dice, 155
    Four Corners, 163
    Power Team Assemble, 123
    Power Up, 121
    Prism Power, 159
    Skewed Corners, 165

dice games (*continued*)
    Solitaire Multi-Match, 115
difference, 47, 56
Digit Disguises, 45
discrete math, 188
divisors, 25
domain vs range, 108
Domoradzki, Amber, 63, 220
Domoradzki, John, 63, 220
Donne, John, ix, 220
Dudeney, Henry Ernest, 152, 220

E

easy games, value of, 7
education, 7, 199
equations, 85, 112
    vs expressions, 94, 112
    vs inequalities, 94
equipment
    cards, 13
    dice, 15
    gameboards, 12, 14
    making a game box, 16
    manipulatives, 50, 116
    tokens, 15
equivalence, 100
Ernest, James, 9, 220
evaluate vs solve, 112
Eves, Howard, 197, 221
Exhaust the Relationships, 116
Exponent Number Train, 125
Exponent Pickle, 127
exponential growth, 120, 124, 131

F

Factor Blaster, 28
Factor Game, The, 28
factors, 25
Factors and Multiples, 23
Factors and Multiples Cooperative, 24
Fermat's Last Theorem, 197
Fezzik vs Inigo, 177
Fight for the Center, 66
Finkel, Dan, 19, 221
first quadrant, 164, 170
Four Corners, 163

Four in a Row, 79
Fractionator, 38
fractions
    bar as grouping symbol, 105
    benchmark numbers, 37
    division of, 39
    negative, 54
Fulton, Brad, 28, 82, 135, 221
Function Machine, 109
functions, 108
fundamental counting principle, 189

G

Gale, David, 178, 221
Galilei, Galileo, 160, 200, 221
gameboards, 12, 14
Games for Math, 7, 224
games, about
    basic rules, 206
    benefits of, 3, 5
    grade level, 7
    house rules, 11
    learn by playing, 11
    modifying, 9
    supply list, 16
games, complete list of
    Algebra Match, 113
    Algebra Multi-Match, 114
    Algebra-Tac-Toe, 115
    Area Battle, 156
    Area Block, 153
    Area Dice, 155
    Consecutive Capture, 53
    Coordinate Gomoku, 171
    Decimator, 39
    Digit Disguises, 45
    Exhaust the Relationships, 116
    Exponent Number Train, 125
    Exponent Pickle, 127
    Factor Blaster, 28
    Factor Game, The, 28
    Factors and Multiples, 23
    Factors and Multiples Cooperative, 24
    Fight for the Center, 66
    Four Corners, 163

games, complete list of (*continued*)
- Four in a Row, 79
- Fractionator, 38
- Function Machine, 109
- Golomb's Game, 151
- Great Escape I, 29
- Great Escape II, 32
- Greater Than, 93
- Grid Fight, 71
- Hidden Hexagon, 166
- Hit Me, 59
- Honeycomb, 73
- Integer Solitaire, 64
- Integer War, 76
- Keeper of the Code, 48
- Krypto, 132
- Krypto Insanity, 136
- Linear War, 173
- My Nearest Neighbor, 35
- Number Riddles, 43
- Number That Must Not Be Named, The, 138
- Number Yoga, 134
- Operations, 103
- Pathways, 81
- Penterimeter, 149
- Perimeter Battle, 157
- Perimeter Block, 155
- Power Krypto, 134
- Power Team Assemble, 123
- Power Up, 121
- Prism Power, 159
- Racetrack, 180
- Radar, 176
- Radian Nim, 178
- Radian Race, 178
- Road Trip Krypto, 134
- Skewed Corners, 165
- Solitaire Multi-Match, 115
- Strike It Out, 62
- Substitution Game, 99
- Tax Collector, 27
- What Two Numbers?, 91
- What's My Rule?, 41
- Year Game, The, 135

Gardner, Martin, 178, 184, 212, 221
Gauss, Carl Friedrich, 20, 222
G'Day Math, 203, 229
geometry, 141
- angles, 143, 169, 179
- area, 70, 150
- Cavalieri's principle, 160
- circles, 179
- coordinate graph, 164, 167, 170, 174, 179, 182
- perimeter, 150
- pi, 179
- polar coordinates, 179, 182
- polygons, 142, 154, 164, 168
- prisms, 160
- quadrilaterals, 164
- radians, 179
- regular vs irregular, 154
- right vs oblique, 160
- vertices, 142, 168
- vocabulary, 141
- volume, 160

Global Math Project, 57, 203, 229
Golden, John, ix, 6, 39, 72, 75, 82, 97, 123, 130, 152, 155, 158, 162, 165, 175, 203, 222
Golomb, Solomon W., 152, 223
Golomb's Game, 151
Gomoku, 171
grade level, 7
graph paper games
- Area Battle, 156
- Area Block, 153
- Area Dice, 155
- Coordinate Gomoku, 171
- Four Corners, 163
- Golomb's Game, 151
- Grid Fight, 71
- Hidden Hexagon, 166
- Honeycomb, 73
- Linear War, 173
- Penterimeter, 149
- Perimeter Battle, 157
- Perimeter Block, 155
- Racetrack, 180

graph paper games (*continued*)
    Radar, 176
    Radian Nim, 178
    Radian Race, 178
    Skewed Corners, 165
graphing
    Cartesian coordinates, 164
    intercepts, 174
    linear equations, 167, 174
    origin, 170
    polar coordinates, 179, 182
    quadrants, 164, 170
    slope, 174
    vectors, 182
Great Escape I, 29
Great Escape II, 32
Great Escape, The, 31, 218
Greater Than, 93
Greene, Joshua, 37, 223
Grid Fight, 71
group games
    and idle time, 206
    Digit Disguises, 45
    Exhaust the Relationships, 116
    Exponent Pickle, 127
    Function Machine, 109
    Hit Me, 59
    Keeper of the Code, 48
    Krypto, 132
    Krypto Insanity, 136
    My Nearest Neighbor, 35
    Number Riddles, 43
    Number That Must Not Be Named, 138
    Number Yoga, 134
    Operations, 103
    Power Krypto, 134
    Power Team Assemble, 123
    Prism Power, 159
    Road Trip Krypto, 134
    Substitution Game, 99
    What's My Rule?, 41
    Year Game, The, 135
growing old, 13

H
Hacker, Andrew, 199, 223
Haines, Kent, 65, 223
Hamilton, Gordon, 202, 215, 223
Hamkins, Joel David, 139, 224
Hands-On Equations, 85
hands-on manipulatives, 50, 116
hexagons, 168
Hidden Hexagon, 166
Hit Me, 59
homeschooling
    adults play along, 8
    allow kids to reason, 187
    common sense, 199, 202
    conversation, 8, 55
    excuse not to think, 50
    goals and temptations, 9
    high school math war, 78
    hints and tricks, 198
    kids think differently, 193
    math club, vii, ix, 9, 61
    math is social, 22
    mistakes, 83, 100, 112, 190, 193, 197
    notice and wonder, 19, 21, 83, 144
    oral proof, 195
    playground of ideas, 141
    slow down, 55
    using games, 7
    value of struggle, 19, 50, 187, 197
    wait time, 191, 194
    word algebra, 90
    word wall, 44
Honeycomb, 73
Hotel Snap, 162, 226
hundred chart games
    Factor Blaster, 28
    Factor Game, The, 28
    Factors and Multiples, 23
    Factors and Multiples Cooperative, 24
    Great Escape I, 29
    Great Escape II, 32
    Tax Collector, 27

I

identities, 112
imagination, 83
inequalities, 94, 112
Inigo vs Fezzik, 177
Integer Solitaire, 64
Integer War, 76
integers, 49, 60
    absolute value, 54, 60
    addition and subtraction, 55
    as exponents, 131
    as vectors, 54
    multiplication and division, 56
    negative times negative, 50, 57
    negative vs minus, 51
inverse operations, 55, 57, 106

J

joy of math, 3, 141

K

K–12 education, 7, 199
Kamii, Constance, 9, 224
Kawas, Terry, 28, 169, 224
Kaye, Peggy, 7, 224
Keeper of the Code, 48
Kilpatrick, William Heard, 199, 225
Krypto, 132
Krypto Insanity, 136

L

Laib, Jenna, 78, 202, 225
Leo, Lucinda, 9, 225
Let's Play Math, 78, 203, 221, 222
linear equations, 167, 174
Linear War, 173
listening to kids, 8, 55, 191
Lombard, Bill, 28, 82, 135, 225
Lyon, Jack, 34, 225

M

magnitude, 54, 60, 182
Making Your Own Sense, 48, 219
manipulatives, 50, 116
math anxiety, 6, 203
math club/math circle, vii, ix, 9, 61
Math Hombre Games, 72, 97, 123, 203, 222

math is boring, 187
Math Is Fun, 182, 227
Math Pickle, 40, 202, 214, 224
mathematicians play games, 10, 22
mean, 68
median, 68
Meyer, Dan, 184, 225
Minecraft, 165
minus vs negative, 51
mistakes, 83, 100, 112, 190, 193, 197
mode, 68
modifying games, 9
Moses, Bob, 201, 226
multiples, 25
multiplication, 56, 70
Multiplication & Fractions: Math Games for Tough Topics, 37
My Nearest Neighbor, 35
mystery, 19

N

natural numbers, 20
nature walk, math as a, 144
negative exponents, 131
negative numbers, 49
negative times negative, 50, 57
negative vs minus, 51
New York Times, The, 199, 218, 223
Nguyen, Fawn, 162, 202, 226
notice and wonder, 19, 21, 83, 144
Nrich Maths, 10, 24, 63, 172, 203, 214, 226
Number Boxes, 78
number line, 51, 164, 170
Number Riddles, 43
Number That Must Not Be Named, 138
number theory, 20
Number Yoga, 134

O

Operations, 103
order of operations, 100, 105
ordered pair, 108, 164

P

parenting
    adults play along, 8

parenting (*continued*)
    finding the right game, 11
    kids think differently, 193
    listening to kids, 8, 55
    value of not knowing, 19, 197
Parker Brothers, 135
Pathways, 81
pencil and paper games
    Digit Disguises, 45
    Exhaust the Relationships, 116
    Factor Blaster, 28
    Factor Game, The, 28
    Function Machine, 109
    Keeper of the Code, 48
    Number Riddles, 43
    Number Yoga, 134
    Road Trip Krypto, 134
    Strike It Out, 62
    Substitution Game, 99
    Tax Collector, 27
    What's My Rule?, 41
    Year Game, The, 135
pentagram, 143
Penterimeter, 149
pentominoes, 149
perfect teacher, 5, 203
Perimeter Battle, 157
Perimeter Block, 155
Philipp, Randolph, 49, 227
pi, 179
pie rule, 172, 207
Pierce, Rod, 182, 227
Plato, 5, 227
playground of ideas, 141
Playing with Math: Stories from Math Circles, Homeschoolers, and Passionate Teachers, ix, 201, 212, 229
polar coordinates, 179, 182
polite numbers, 20
Polya, George, 199, 227
Post, Sonya, 102, 119, 187, 227
Power Krypto, 134
Power Team Assemble, 123

Power Up, 121
Prealgebra & Geometry Printables, 12, 15
prime number, 26
Princess Bride, The, 177
Prism Power, 159
problem solving
    as a journey, 55, 187, 197
    celebration of, 215
    dreaming time, 197
    finding the sweetness, 3
    imagination and, 83
    not knowing, 19
    value of struggle, 19, 50, 187, 197
    word algebra, 90
problem vs exercise, 197
product, 39, 47
puzzles
    factors and multiples, 24
    hands-on equations, 85
    making sense of integers, 50
    missing-number equations, 56
    playground balance, 84
    polite numbers, 20

Q

quadrilaterals, 164
queen of the sciences, 20
quotient, 47

R

Racetrack, 180
Radar, 176
Radian Nim, 178
Radian Race, 178
radians vs degrees, 179
Radical Equations, 201, 226
Ramblings of a Math Mom, 22, 217
range (of data), 68
range (of function), 108
rate of change, 174
real-life mathematics, 10, 22, 49, 197
reason and imagination, 83
reasoning skills, 6, 187, 197
reflex angle, 143, 169
replication, 73
Road Trip Krypto, 134

rules
- changing, 9
- dealing the cards, 208
- discard pile, 209
- draw pile (stock), 209
- fishing pond, 209
- hand vs round vs game, 208
- house rules, 11
- keeping score, 211
- learn by playing, 11
- misdeal, 209
- number of players, 206
- pie rule, 172, 207
- rotation of play, 208
- shuffle and cut, 207
- who goes first, 207

S

Salomon, Paul, 202, 228
Sawyer, W. W., 83, 85, 228
scale, 52
School of Islamic Geometric Design, 148, 218
Scientific American, 178, 184, 221
secret code, 45
Serra, Michael, 184, 225
Shore, Chris, 58, 228
Simply Great Math Activities, 28, 82, 221, 225
Singh, Marina, 31, 228
Skewed Corners, 165
slope, 174
Slow Reveal Graphs, 202, 225
Smith, Nick, 75
Snake Lady vs Captain Victory, 123
solitaire games
- Exhaust the Relationships, 116
- Exponent Pickle, 127
- Integer Solitaire, 64
- Number Yoga, 134
- Solitaire Multi-Match, 115
- Substitution Game, 99
- Year Game, The, 135

Solitaire Multi-Match, 115
solve vs evaluate, 112

SolveMe Puzzles, 202, 220
Sorooshian, Pam, 10, 214, 228
square numbers, 30
Square One TV, 82, 219
staircase numbers, 21
star polygons, 142
statistics, 68
Steve vs Tony, 95
Steward, Don, 107, 229
Strike It Out, 62
struggle, value of, 19, 50, 187, 197
stump the teacher, 139
substitution, 100, 112
Substitution Game, 99
sugar and medicine, 3
Sullivan, Christine, 37, 229
sum, 47, 56
supplies
- cards, 13
- dice, 15
- gameboards, 12, 14
- making a game box, 16
- manipulatives, 50, 116
- tokens, 15

swap2, 172
symmetry, 195

T

Talking Math with Your Kids, 8, 220
Tanton, James, 57, 70, 202, 203, 212, 214, 229
Tax Collector, 27
teaching
- allow kids to reason, 187
- common sense, 199, 202
- differentiation, 7
- excuse not to think, 50
- games in the classroom, 7
- high school math war, 78
- hints and tricks, 198
- ideal strategy, 5, 203
- kids think differently, 193
- listening to kids, 8, 55
- math is social, 22
- mistakes, 83, 100, 112, 190, 193, 197

teaching (*continued*)
    notice and wonder, 19, 21, 83, 144
    oral proof, 195
    playground of ideas, 141
    slow down, 55
    speed games, 7
    value of struggle, 19, 50, 187, 197
    wait time, 191, 194
    word algebra, 90
    word wall, 44
tokens, types of, 15
Tony vs Steve, 95
tree diagram, 188

U

unary operator, 52
Unsummables, The, 21
useless games, 6

V

VanHattum, Sue, ix, 3, 201, 212, 229
variables, 80, 108
Vector (character), 54
vector (integer), 54
vector (movement), 182
Venn diagrams, 40
vertex (vertices), 142, 168
Visual Patterns, 202, 215, 226

W

Way, Jenni, 7, 229
what mathematicians really do, 10, 22, 197
What Two Numbers?, 91
What's My Rule?, 41
Why is negative times negative positive?, 57
Woods, Martin, 115, 230
World War II, 31, 34

X-Y-Z

Year Game, The, 135
Young Children Reinvent Arithmetic, 9, 224
YouTube, 58, 207, 217, 219, 224, 229
Yovich, Daniel, 135
Zaslavsky, Claudia, 8, 212, 230
zero, 51, 60, 124

About the Author

FOR MORE THAN THREE DECADES, Denise Gaskins has helped countless families conquer their fear of math through play. As a math coach and veteran homeschooling mother of five, Denise has taught or tutored at every level from preschool to precalculus. She shares math inspirations, tips, activities, and games on her blog at DeniseGaskins.com.

Denise encourages parents and teachers to look at math with fresh eyes. "We want to explore the adventure of learning math as mental play, the essence of creative problem solving. Mathematics is not just rules and rote memory. Math is a game, playing with ideas."

Get More Playful Math

Join Denise's free newsletter list to receive an eight-week "Playful Math for Families" email course with math tips, activities, games, and book excerpts. Also, about once a month, she'll send out additional ideas for playing math with your kids. And you'll be one of the first to hear about new math books, revisions, and sales or other promotions.
tabletopacademy.net/mathnews

Math Games, Tips, and Activity Ideas for Families.

TabletopAcademy.net/MathNews

Books by Denise Gaskins

tabletopacademy.net/playful-math-books

"Denise has gathered up a treasure trove of living math resources for busy parents. If you've ever struggled to see how to make math come alive beyond your math curriculum (or if you've ever considered teaching math without a curriculum), you'll want to check out this book."
— Kate Snow, author of Multiplication Facts That Stick

Let's Play Math: How Families Can Learn Math Together — and Enjoy It

Transform your child's experience of math!

Even if you struggled with mathematics in school, you can help your children enjoy learning and prepare them for academic success.

Author Denise Gaskins makes it easy with this mixture of math games, low-prep project ideas, and inspiring coffee-chat advice from a veteran homeschooling mother of five. Drawing on more than thirty years of teaching experience, Gaskins provides helpful tips for parents with kids from preschool to high school, whether your children learn at home or attend a traditional classroom.

Don't let your children suffer from the epidemic of math anxiety. Pick up a copy of *Let's Play Math,* and start enjoying math today.

The *Math You Can Play* Series

You'll love these math games because they give your child a strong foundation for mathematical success.

By playing these games, you strengthen your child's intuitive understanding of numbers and build problem-solving strategies. Mastering a math game can be hard work. But kids do it willingly because it's fun.

Math games prevent math anxiety. Games pump up your child's mental muscles, reduce the fear of failure, and generate a positive attitude toward mathematics.

So what are you waiting for? Clear off a table, grab a deck of cards, and let's play some math.

The *Playful Math Singles* Series

The *Playful Math Singles* from Tabletop Academy Press are short, topical books featuring clear explanations and ready-to-play activities.

312 Things To Do with a Math Journal includes number play prompts, games, math art, story problems, mini-essays, geometry investigations, brainteasers, number patterns, research projects for all ages.

70+ Things To Do with a Hundred Chart shows you how to take your child on a mathematical adventure through playful, practical activities. Who knew math could be so much fun?

More titles coming soon. Watch for them at your favorite bookstore.

www.ingramcontent.com/pod-product-compliance
Lightning Source LLC
Chambersburg PA
CBHW021143080526
44588CB00008B/186